Responses in the Yield of Milk Constituents to the Intake of Nutrients by Dairy Cows

Biotechnology and Biological Sciences Research Council
(formerly the Agricultural and Food Research Council)
Technical Committee on Responses to Nutrients
Report no. 11

Reprinted from
Nutrition Abstracts and Reviews
Series B
Volume 68 No. 11

CABI *Publishing*

CABI *Publishing* – **a division of CAB INTERNATIONAL**

CABI *Publishing*
CAB INTERNATIONAL
Wallingford
Oxon OX10 8DE
UK

CABI *Publishing*
10 E. 40th Street,
Suite 3203
New York, NY 10016
USA

Tel: +44 (0)1491 832111
Fax: +44 (0)1491 833508
Email: cabi@cabi.org

Tel: +1 212 481 7018
Fax: +1 212 686 7993
Email: cabi-nao@cabi.org

A catalogue record for this book is available from the British Library, London, UK.
A catalogue record for this book is available from the Library of Congress, Washington DC, USA.

ISBN 0 85199 284 6

Typeset by Solidus (Bristol) Limited
Printed and bound by Biddles Ltd, Guildford and King's Lynn

Contents

List of Tables

List of Figures

Preface

The inadequacy of existing UK energy and protein requirement systems for dairy cattle to predict responses in milk yield and composition to changes in nutrient supply led to the setting up of a Working Party under the auspices of the former AFRC Technical Committee on Responses to Nutrients (TCORN) to produce a report on this important topic. A first draft of their report was written by the members, and the manuscript extensively revised, but not completed, before funds were withdrawn from this area of study.

Further attempts were made by the group to seek alternative sources of support but without success. However, recently CABI *Publishing* expressed an interest in publishing the Dairy Cow Report as a review article in *Nutrition Abstracts and Reviews* and as a paperback reprint. Fortunately, the manuscript of the draft report (complete with figures and tables) still existed as hard copy and as a computer file. Also the MAFF *LINK* Project 'Feed into Milk' agreed to support the editorial work necessary to finalize the report.

The Chairman and Technical Secretary consulted the other members of the Working Party, and they agreed to collaborate with Mr G. Alderman, appointed as Editor, to update and bring the manuscript to a state ready for publication. All the work was done by post, with no formal meetings of the Working Party having taken place to approve the final draft chapters.

Joint funding by CABI *Publishing* and the MAFF *LINK* Project 'Feed into Milk' arranged for the completion of this Report is hereby gratefully acknowledged.

Nutrition Abstracts and Reviews (Series B)
BIOTECHNOLOGY AND BIOLOGICAL SCIENCES RESEARCH COUNCIL
Polaris House, North Star Avenue, Swindon SN2 1UH
(formerly the Agricultural and Food Research Council)
This publication was prepared by the Working Party on Responses in the Yields of Milk Constituents by Dairy Cows with particular reference to the Supply of Nutrients and their Metabolism

Membership

Professor P.C. Thomas	The Scottish Agricultural College, Central Office, West Mains Road, Edinburgh EH9 3JG, UK (Chairman).
Professor D.E. Beever	Centre for Dairy Research, Department of Agriculture, University of Reading, Earley Gate, Whiteknights, Reading RG6 6AT, UK.
Professor P.J. Buttery	University of Nottingham, School of Agriculture, Sutton Bonington, Loughborough, Leicestershire LE12 5RD, UK.
Professor J.C. MacRae	Rowett Research Institute, Bucksburn, Aberdeen AB21 9SB, UK.
Professor J.D. Oldham	The Scottish Agricultural College, Bush Estate, Penicuik, Midlothian EH26 0QE, UK.
Professor C. Thomas	The Scottish Agricultural College, Auchincruive, Ayr KA6 5HW, UK (Technical Secretary).

Terms of reference

To formulate new proposals for the prediction of responses in the yield of milk constituents by dairy cows with particular reference to the supply of nutrients and their metabolism.

Abbreviations

A–V	arterio-venous
AA	amino acid
AAT	amino acids truly absorbed in the small intestine
ADPLS	apparently digestible protein leaving the stomach
AP	available protein
ATP	adenosine triphosphate
CHO	carbohydrate
CNCPS	Cornell net protein and carbohydrate system
CS	condition score
DC	degradable carbohydrate
DCF	digestible crude fibre
DCP	digestible crude protein
DE	digestible energy
DM	dry matter
DMI	dry matter intake
DNFE	digestible N-free extract
DOM	digestible organic matter
dsi	digestibility of the UDP fraction
DVE	true protein digested in the small intestine
EFN	endogenous faecal nitrogen
FME	fermentable metabolizable energy
FOM	fermentable organic matter
k_d	rate of degradation
k_p	solid outflow rate
LCFA	long-chain fatty acids
MAFF	Ministry of Agriculture, Fisheries and Food
MCP	microbial crude protein
ME	metabolizable energy
MFN	metabolic faecal nitrogen
MP	metabolizable protein
NADH	nicotinamide adenine dinucleotide

NAN	non-ammonia nitrogen
NDF	neutral detergent fibre
NE	net energy
NPN	non-protein nitrogen
PAH	para-aminohippuric acid
PDI	protein digested in the intestine
QDP	quickly degraded protein
RDN	rumen degradable nitrogen
RDP	rumen degradable protein
RV	relative value
SDP	slowly degraded protein
SE	starch equivalent
TDN	total digestible nutrients
UDP	undegraded dietary protein
VFA	volatile fatty acids
W	live weight

1. Summary

This report considers the energy and protein feeding systems for dairy cows currently used in Britain, their limitations and their relevance to the needs of the industry. It then examines the principles underlying alternative systems embodying concepts of nutrient supply and nutrient use and considers the scientific information which will be needed for their development as viable alternatives to the current systems.

Current systems of diet formulation adopted for use in the UK, based on the ARC metabolizable energy (ME) and AFRC metabolizable protein (MP) proposals, use factorial assessments of the cow's nutrient requirements to calculate the supply of energy and protein that must be provided in the diet to meet those requirements. There is little or no consideration of the animal's response to changes in the nutrient supply. Furthermore, the systems do not satisfactorily accommodate interactions between dietary constituents in their effects on digestion and metabolism, with only limited allowance for energy–protein interactions in the rumen. Lastly, the systems fail to comment on the effects of diet on the partition of nutrient use between milk and body, and on the individual yields of milk fat, protein and lactose. This is an important limitation, since the relative amounts of the individual milk constituents that are produced determine the commercial value of the milk. Thus the argument presented is not that current systems are inaccurate but that they lack relevance to the future needs of milk producers.

The Working Party considers that a new feeding system for dairy cows should aim to predict: voluntary feed intake; the partition of nutrient use between milk production and tissue deposition; and the yield of milk fat, protein and lactose in relation to both short- and long-term effects of nutrition.

A fundamental change in approach is recommended which will allow nutrient supply from the diet and nutrient utilization in the cow to be described in more relevant biological terms than those used at present. This will allow interactions in both digestion and metabolism between protein (amino acids) and energy-yielding substances to be described more satisfactorily and will permit the distinctive influences of the individual

1

energy-yielding substances on nutrient utilization to be taken into account. The physical and biological characteristics of the individual cow must also be explicitly recognized and incorporated in appropriate terms into the system for response prediction.

It is proposed that a system to predict production of milk constituents from the amount, chemical composition and nutritional characteristics of the cow's diet could be developed in two parts: (i) a digestion model, which relates diet to the main nutrient products absorbed by the animal from the digestive tract, and (ii) a utilization model, which describes the absorption and metabolism of the nutrients. The system would be constructed in the form of a computer model, with inputs describing the feeds, the genotype and physiological condition of the cow, and the environment.

With current knowledge it should be possible to develop a model to predict, quantitatively, the main end-products of digestion in the rumen, small intestine, caecum and colon of the cow. These end-products (acetate, propionate, butyrate, glucose, amino acids (AA) and long-chain fatty acids (LCFA)) form the nutrient supply to the animal. Development of the model to a level suitable for practical application will require improved chemical and nutritional characterization of feeds and a better understanding of some aspects of the digestive process, particularly the rumen.

With present knowledge it is not possible to translate estimates of nutrient supply into terms reflecting the substrates available to the individual body tissues in other than a crude and simplistic way. However, there is a realistic prospect for the development of a predictive model based on a simple representation of nutrient utilization which is biologically sound and which incorporates terms reflecting the main features of the cow's metabolism. The general approach adopted is to consider nutrient use for: basal metabolism; fetal growth; milk fat, protein and lactose synthesis; and body protein and fat deposition. Basal metabolism and fetal growth are regarded as obligatory processes with a first charge on the nutrient supply, whilst the synthesis of the milk constituents and of body protein and fat are regarded as responsive functions, varying with nutrient supply and competing with each other for available nutrients as lactation progresses.

Improved characterization of the cow is an essential feature of the modelling approach to response prediction which is advocated. The prediction model must include terms reflecting the cow's potential for the synthesis of milk constituents, the synthesis or metabolization of fat and body protein, and must recognize the cow's current biological status in relation to those potentials.

On the basis of the information presented, the Working Party concludes that the development of a feeding system that will allow the yields of milk fat, protein and lactose to be predicted from a knowledge of the nutrient supply furnished by the diet and the metabolism of those nutrients by the cow should be encouraged. However, the development of a satisfactory system for practical application will require substantial and directed research effort. Recommendations are made for future research on: the characterization of feedstuffs; selected aspects of digestion and nutrient supply; the utilization of substrates by the cow; and cow characterization. These recommendations are designed specifically to facilitate the development of an appropriate predictive system.

2. Introduction

As a result of recent trends in the markets for milk and milk products, the dairy industry is undergoing a period of unprecedented change. Economic pressures have increased with the introduction of quotas on milk production, and there has been a continued narrowing in the differential between the cost of milk production and the price of milk.

At present approximately half of the cost of milk production can be attributed to feed costs, emphasizing the need for accurate feeding systems to guide diet formulation and the allocation of feed resources, and to facilitate management decisions on feed inputs relative to milk outputs. Moreover, within the modern context, such systems must have regard to the compositional quality of milk and the need to regulate milk composition and the production of milk constituents to meet changing market circumstances.

This report briefly considers the current energy and protein requirement systems for dairy cows currently used in Britain, their limitations and their relevance to the needs of the industry. It then examines the principles underlying alternative systems embodying concepts of nutrient supply and nutrient use and considers the scientific information which will be needed for their development as viable alternatives to the current systems.

During the preparation of the report several major areas of ruminant nutrition were reviewed and the conclusions of those reviews are presented in summary form. However, it was not the purpose of the report to provide a comprehensive documentation of the literature and only key references or relevant review articles are referred to in the text.

3. Current Feeding Requirement Systems and their Limitations

3.1 The current systems and their development

3.1.1 Energy requirement systems

For the major part of this century the energy system adopted for use in Britain was the starch equivalent (SE) system initially developed in the early 1900s by Oskar Kellner at the Mockern Laboratory in Germany (Kellner, 1905). The system involved the description of the cow's requirements for maintenance, fattening and lactation in net energy (NE) terms and the allocation of NE values to individual feeds. Rather than using the calorie energy unit, Kellner chose to express both requirements and feed values as weights by using the NE value of 1 kg of starch as a standard. He recognized that the NE values of feeds would differ according to whether they were being used for maintenance, fattening or milk production. But instead of ascribing variable NE values to feeds, he adopted the NE for fattening as a fixed standard and modified his estimates of NE requirements to express them in corresponding terms. For this he applied corrections based on estimated efficiencies of utilization of energy for fattening:maintenance:lactation of 1:1.3:1.25.

Kellner's NE system provided the background to the standard allowances of SE for maintenance and milk production which were proposed in Britain by T.B. Wood in *Rations for Livestock* (MAFF, 1921). Over the following years the main tenets of this approach were accepted largely without challenge although there was much debate about the appropriate allowances of SE for milk production. Values were raised from the initial 0.22 kg kg^{-1} milk produced to 0.26 kg kg^{-1} (MAFF, 1960), culminating in the proposal by Blaxter (1959) of a value of 0.29 kg kg^{-1}, which was designed to include a safety margin to ensure that no more than one cow in 20 was underfed when the standard was applied.

Major deficiencies in the SE system were progressively recognized, nonetheless (ARC, 1965, 1980). The assumption that the NE value of a feed was constant was invalidated by the finding that the NE contents for fattening of feeds declined with increasing level of feeding. Similarly, the assumption that the NE values of feeds for the separate functions of maintenance, fattening and lactation were related by a simple ratio was progressively disproved by calorimetric experiments on the effects of feed composition on the partial efficiencies of energy utilization. Recognizing the shortcomings of the SE system, the ARC Technical Committee on the Nutrient Requirements of Farm Livestock (ARC, 1965) proposed a change to the ME system, developed by the late Sir Kenneth Blaxter (Blaxter, 1962). The important features of the new system were:

1. Inclusion of a correction to allow for reduction in availability of dietary ME with increasing levels of feeding above maintenance.
2. Use of separate partial efficiencies of utilization of ME for maintenance (k_m), fattening (k_f) and lactation (k_l), calculated from the metabolizability of dietary gross energy (q_m).

In comparison with the earlier SE standards the ME system advocated reduced allowances of dietary energy at low levels of milk production and increased allowances at high levels. The accuracy with which the ME system and the SE system predicted milk production for animals fed approximately to theoretical requirements was examined at an early stage by a Ministry of Agriculture, Fisheries and Food (MAFF) Interdepartmental Working Party (MAFF, 1972; Alderman *et al.*, 1973). It concluded that the proposed ME system was an improvement over the SE system although this was by no means clearly evident from the results provided in the Working Party's report. Large deviations were noted between predicted and observed milk yields with both systems, particularly over short periods of time where the recorded value of live weight change might be expected to have a large error of measurement. However, in all sets of data examined there was evidence that the ME system underestimated predicted yields at low levels of production and overestimated them at high levels. It was not possible to decide whether these discrepancies arose through a failure to estimate adequately the ME values of the feeds, corrections for feeding level on realized dietary ME or through errors in the estimates of maintenance requirement, the efficiency of utilization of ME for milk or the energy deposited in association with body weight gain; both types of error must be suspected. Nevertheless, it was agreed, on the basis of increased flexibility and ability to incorporate new data that the ME system should be adopted after suitable modifications for practical use. The modifications entailed disregarding the effect of level of feeding on ME supply and adopting constant efficiencies of utilization of ME for maintenance and lactation. Thus the modified ME system enshrined in MAFF Technical Bulletin 33 (MAFF, 1975), in the interests of simplification judged necessary to assist its adoption in practice, failed to incorporate two of the main elements of the ME system as originally proposed.

Further alterations to the ARC (1965) ME system were advocated by ARC (1980), based on the extensive work on the energy requirements of lactating dairy cows by Van Es at Lelystad and by Flatt, Moe and Tyrrell at USDA, Beltsville (Van Es *et al.*, 1970; Moe *et al.*, 1972; Van Es, 1978). Terms were included to allow for energy released from or deposited in body tissues of lactating cows, and it was also recognized that tissue energy gain was more efficient in the lactating rather than in the non-lactating animal. These changes resulted in the revision of the standards to provide for an increase of approximately 5% in calculated ME requirements at low milk yields and a corresponding reduction of approximately 4% at high yields. An Advisory Services Working Party was set up in 1982 to consider and test the ARC (1980) recommendations for energy requirements of ruminants. Their report (AFRC, 1990) found that, as far as dairy cattle were concerned, the recommended ME system was superior to the simplified system of MAFF (1975) in its ability to account for milk yields, although there were stages of lactation effects, with under-estimation being greater in early lactation (see Section 3.2.1). Since the need to have a simplified ME system no longer existed, with the widespread availability of computers to professional ruminant nutritionists, adoption of the ARC (1980) proposals was recommended for the ME requirements of dairy cows, with a safety margin of 5% added. These recommendations were accepted in 1992 and implemented in the UK by means of an advisory manual (AFRC, 1993).

3.1.2 Protein requirement systems

Replacement of the traditional and much criticised digestible crude protein (DCP) system (Evans, 1960) with an improved system based on concepts of 'available protein' was recommended by ARC (1965). However, the recommendation was not adopted by MAFF because of difficulties in defining acceptable practical standards and minimum requirements for available protein digestion in the ruminant (MAFF, 1972). So it was not until publication of the ARC 'protein system' (ARC, 1980) that a real alternative to the DCP system became available. In their considerations the ARC (1980) Working Party addressed themselves to the failure of existing protein rationing systems to describe adequately the differing nutritional values of feeds of corresponding DCP content and to pay regard to the influences of dietary energy supply on protein utilization. Their proposals were to describe feeds in terms of their rumen degradable protein (RDP) and undegraded dietary protein (UDP) contents and to relate the microbial capture of RDP in the rumen to dietary ME supply. Thus it was possible to define protein supply to the animal in terms of the microbial protein and UDP digested and absorbed in the small intestine, and to link the supply of absorbed protein with the corresponding 'tissue' protein requirements for maintenance, tissue synthesis and milk production.

Over the period of preparation of the ARC (1980) report, research on protein digestion had advanced rapidly, however, and a supplementary report was therefore commissioned to take account of new findings. This report (ARC, 1984) presented a revised version of the ARC protein system incorporating a number of changes in concept as well as detail. Thus the protein degradability of a feed was no longer assumed to be described by a single value but was estimated from the potential degradability, the rate of degradation of the feed, and the residence time of feed particles in the rumen, which in turn depended on the level of feeding. Capture of RDP was, as before, considered to be related to dietary ME supply but the relationship was now regarded as varying to some degree with level of feeding (as multiples of maintenance) rather than being constant. There were increases in the values for total endogenous nitrogen loss and in the efficiencies of utilization of absorbed protein, and these were linked with a change from apparent to true digestibility values for the digestion of protein in the small intestine. Finally, there was recognition of the importance of the AA composition of the protein absorbed. It was suggested that, particularly for dairy cows, improvements in the prediction of responses to protein supply might be achieved if estimates of requirement and supply were expressed in terms of essential AAs rather than total AAs.

A number of broadly similar ruminant protein systems have also been constructed and published:

- INRA (1978, 1988): protein digested in the intestine (PDI);
- NRC (1985): available protein (AP);
- Madsen (1985): AA truly absorbed in the small intestine (AAT);
- SCA (1990): apparently digestible protein leaving the stomach (ADPLS);
- Tamminga *et al.* (1994): true protein digested in the small intestine (DVE).

The main difference between these systems lies in the choice of diet parameter used to predict microbial protein synthesis, i.e. fermentable organic matter (FOM) for PDI, total digestible nutrients (TDN) for AP, digestible carbohydrate (DCF + DNFE from Weende analyses) for AAT and ME for ADPLS. FOM was calculated from DOM corrected for oils and fats, fermentation acids and undegraded protein. In the case of the Dutch DVE unit, FOM was also corrected for undegradable starch. Microbial yield in all systems, except ADPLS, was in a fixed ratio to the chosen parameter describing fermentable nutrients. All systems adopt broadly similar digestibilities of microbial true protein, and all (except the DVE system) use fixed efficiencies of utilization of the absorbed microbial and feed amino acids for the various metabolic functions.

The other main difference between systems was the method of dealing with metabolic faecal nitrogen (MFN), which Jones *et al.* (1996) showed was responsible for the major part of the differences in calculated MP requirements from the various systems. INRA (1978, 1988) excluded MFN entirely, since maintenance MP requirements were assessed from feeding trials, whilst ARC (1984), by implication, included MFN in the maintenance requirement, but did not carry this over to the total diet dry matter intake (DMI). The AP and AAT systems and the CAMDAIRY model (Hulme *et al.*, 1986) all include MFN as an obligatory AA requirement, relating it either to DMI or, in the case of NRC (1985), to indigestible DMI. In the case of the Australian ADPLS system, a term endogenous faecal N (EFN) was adopted, numerically similar to the MFN of NRC (1985), but related to DMI. Tamminga *et al.* (1994) decided to follow the logic of MFN being a feed-related quantity (even if an obligatory AA expense of the animal) and deduct their estimate of MFN per kilogram of feed dry matter from the calculated feed dry matter DVE value, which is based on the estimated FOM supply from the feed in question. The view that MFN is only a component of maintenance AA-N requirement, not a feed intake-related parameter, leads to significant differences in the calculated MP requirements between AFRC (1992), INRA (1988) and all the other MP systems.

Some aspects of the ARC (1984) system were less than satisfactory. Diets formulated in accordance with its proposals were found to be less than optimal for milk production, values of only 14% crude protein in the dietary dry matter being calculated, compared with current recommendations and practice (from the DCP system) of about 16% (Alderman, 1987). There was also a need for further improvement in the assessment of protein degradability and in the prediction of its variation in response to changes in the functional activity of the rumen microbes and the digestive physiology of the animal. Also, though the system now recognized some variability in microbial growth yield, i.e. microbial protein yield from the rumen per unit of ME, it did not provide an adequate basis from which the appropriate microbial growth yield can be calculated. Unique values were recommended to apply to well balanced mixtures of forage and concentrate feeds, to wholly concentrate diets, and to diets of grass silage or combinations of grass silage and concentrates. These values imply discontinuity in the microbial growth function, which seem unlikely to be biologically correct. Variation in microbial growth yield in dairy cows compared with standardized estimates (as in ARC, 1980, 1984) has been identified as a major limitation of the ability of the systems to predict the responses of dairy cows to changes in protein feeding (Oldham, 1984). In order properly to account for recognized deviations in microbial growth yield even within the broad classes of diets identified in ARC (1984), it will be necessary to identify continuous functions which relate measurable characteristics of feeds and of animals to the availability of rumen degradable nitrogen (RDN) and its capture by rumen microbes. With many diets, organic matter digestibility increases as dietary protein content is raised (Gordon, 1979) and this appears to be due to a change in microbial activity in the rumen. In order to predict such responses it is necessary to improve the description of microbial growth functions.

Additionally, whilst both the apparent and true absorption coefficients for microbial AAs in the small intestine appear to be predictable and relatively constant, the same does not seem to be true for the digestibility of the UDP fraction (*dsi*). Some attempts have been made to predict *dsi* (Wilson and Strachan, 1981; Webster *et al.*, 1984), but INRA (1988) rely upon estimates of *dsi* by statistical analysis of faecal N into its microbial, endogenous and feed components. The DVE system of Tamminga *et al.* (1994) uses the *in vivo* mobile nylon bag technique of Van Straalen and Tamminga (1990). This reliance upon *in vivo* estimates is likely to be resolved by an improved characterization of the diet rather than by a change in concept.

In contrast, the assumption of a high and constant efficiency of utilization of absorbed AAs requires a change in concept. High efficiencies may be observed only when the AA profile of the absorbed protein is ideal or close to it, and AA supply is limiting. Thus it may be argued that the net efficiency with which absorbed AAs are used for milk protein production (k_{nl}) should be regarded not as a constant but as a variable to be predicted within the feeding system from attributes both of the diet (balance of AAs supplied) and the cow (efficiency of use of ideal AA mixture).

A number of these issues were addressed by a joint advisory services Working Party set up in 1982 to test the ARC 1980/84 proposals, modify them if necessary and recommend a protein requirement system for ruminants for adoption in the UK. Their report (AFRC, 1992) proposed a metabolizable protein (MP) system, including a new unit, fermentable metabolizable energy (FME), calculated from the feed or diet ME by discounting the gross energy from oils, fats and fermentation acids, because they are not sources of energy for rumen bacteria. Microbial yield per unit of FME and solid outflow rate were curvilinearly related to level of feeding. The equation proposed by Ørskov and McDonald (1979) was used to calculate both the type (quickly and slowly degraded, QDP and SDP) and amounts of protein degraded in the rumen. The proportion of nucleic acids in microbial crude protein was reduced to 0.25 based on an EEC ring test (AFRC, 1992). Whilst the ARC (1984) recommended value of 0.85 was retained for the efficiency of utilization of an *ideal protein*, lower fixed efficiencies of AA utilization (k_n) were allocated to the different physiological functions: lactation, growth, pregnancy and fibre growth. This was done by the introduction of the term relative value (RV) instead of biological value, which incorporated the role of AA composition of the absorbed MP in influencing efficiency of utilization. The recommendations were accepted for adoption in the UK in July 1992 and implemented by an advisory manual (AFRC, 1993).

The Dutch metabolizable protein (DVE) system (Tamminga *et al.*, 1994) refined the INRA FOM term by deducting estimates of rumen resistant (undegraded) starch, and measured feed UDP digestibility directly by using a mobile nylon bag containing the residues from an *in situ* dacron bag degradability which had been suspended in the rumen for 72 h or more (Van Straalen and Tamminga, 1990). Recent work (Antoniewicz *et al.*, 1998) has shown that a pepsin/pancreatin digestion of mobile dacron bag residues can give comparable values to those obtained from the *in vivo* mobile nylon bag technique.

Subnel *et al.* (1994) also proposed that the efficiency of DVE (MP) utilization (k_{nl}) should be a function of MP/ME, varying from 0.5 to 0.7 as the ratio approached optimum. This approach agrees with the findings of AFRC (1992) and Webster (1992), who showed that increments of dietary MP without addition of ME resulted in a marginal response of 0.2 in milk net protein. Thus the Dutch DVE protein system is the first to incorporate a nutritional response into its protein requirement model. Oldham (1996) has subsequently attempted an explanation of this low response to additional MP. If absorbed AAs are used as general purpose energy (ME), an efficiency of utilization of 0.33 can be calculated, whereas if used to synthesize lactose, the value is 0.28.

3.1.3   Integrated energy and protein systems

A spreadsheet model of the microbial digestion and metabolism in the rumen of dairy and beef cattle, the Cornell net protein and carbohydrate system (CNCPS) was published in 1992 by the research team at Cornell University (Fox *et al.*, 1992; Russell *et al.*, 1992; Sniffen *et al.*, 1992; O'Connor *et al.*, 1993). It relies upon NRC (1985) and NRC (1988), with minor modifications, for estimates of the energy and protein requirements of the dairy cow under study. The adequacy of a specified diet in meeting the set nutritional targets is assessed, together with much information about conditions in the rumen, the balance of the fermentation and how these may be optimized.

The nutrient fractions in feeds are divided into two broad groups, cell contents and cell walls as in the Goering and Van Soest (1970) analytical procedures, resulting in five carbohydrate (CHO) fractions and six protein fractions, each of which is assigned a rate of degradation in the rumen (k_d), which is not modified by the estimated solid outflow rate (k_p). The latter only affects the residence time of feed particles in the rumen, which influences the amount of nutrient degraded at the specified fixed rate of degradation.

The microbial population is simplified down to only two groups, cell wall (structural CHO)-fermenting bacteria (SC) which only utilize ammonia as their N substrate, and non-cell wall (non-structural CHO)-fermenting bacteria (NSC), which utilize both ammonia and polypeptides. The influence of protozoa is only introduced as a cause of a reduction in maximum predicted microbial yield. Only fermentable CHO and peptides can contribute to microbial growth, and their N supply is assumed not to be limiting, rather as in the Scandinavian AAT (Madsen, 1985) and Dutch DVE (Tamminga *et al.*, 1994) protein models, i.e. the rumen N balance is computed and then assessed for adequacy. Microbial yield is predicted by the Pirt (1965) double reciprocal equation, which requires estimates both of microbial maintenance requirements and the maximum microbial yield. Microbial yield is also influenced by the rumen pH predicted from a parameter, effective neutral detergent fibre (eNDF), which combines feed particle size measurements with NDF estimates (Mertens, 1985). Originally, no estimates of volatile fatty acid (VFA) production, VFA proportions or methane production were made, although the subsequent conversion in the model from estimated total digestible nutrients (TDN) via DE to ME takes account of the energy implications of methane production. Later papers from the same group (Pitt *et al.*, 1996) have attempted to remedy this important omission.

The model computes the amounts of AAs, both microbial and feed, leaving the rumen and absorbed in the intestines, largely along the lines of the NRC (1985) AP system. Only the absorbed AAs are taken into detailed account in an AA sub-model (O'Connor *et al.*, 1993), which predicts the adequacy of the AA supply from the diet in meeting the production targets set, assuming that the AAs absorbed are available unaltered to the productive tissues, as many earlier models have done, e.g. Gill *et al.* (1984). All other absorbed nutrients are combined (with protein) into an estimate of TDN and then converted sequentially to estimates of DE, ME and NE for lactation (NE_l) as in NRC (1988). This means that there are no predicted energy or nutrient interactions upon milk composition in the model. Cow body-weight change is affected by growth to maturity at 4 years of age as defined by Fox *et al.* (1992) and the energy surplus or deficits resulting from meeting the set milk yield targets.

3.2   Accuracy of prediction of current systems

The accuracy of prediction of the energy and protein rationing systems currently used in the UK have been examined by the aforementioned MAFF Interdepartmental Working Parties (AFRC, 1990, 1992). It is not the intention to catalogue their findings here, but it is relevant to consider some of the more important conclusions that have emerged.

3.2.1   Metabolizable energy

The Working Party on Energy (AFRC, 1990) showed that the ME system predicted total energy output, with a bias between observed and predicted output of only +2.5%. However, there was considerable variability reflected in a standard deviation (SD) of the difference between observed and predicted output of ±12%. However, this variability combines the variability in the estimates of feed ME values and animal variability in the utilization of the ME supplied. Thus the current system provides an accurate prediction of total energy output for the 'average cow' but is less satisfactory in its predictions for the individual. The implication of the standard deviation in practice is that relatively large safety margins (increases in requirement) need to be applied to ensure that individual cows are not underfed. Values derived from a statistical analysis similar to that used by Blaxter (1967) are shown in Table 3.1 and indicate that ME allowances should be almost 20% above requirements to ensure that no more than 5% of cows are underfed. In this context the currently applied safety margin of 5% apparently results in 35% of cows being underfed. The above conclusions are, however, only applicable to energy output over a relatively long period of the lactation (25–30 weeks). Over shorter periods (e.g. 4–6 weeks) the errors of prediction, both in terms of bias and SD, increase substantially such that the safety margin of 30% is required to ensure that only 1 in 20 cows is underfed. Similar large errors over successive short periods of lactation have been noted previously and attempts have been made to ascribe the errors to the value assumed for the energy content of the body weight change (Alderman *et al.*, 1982) or to the assumed efficiency of utilization of ME for lactation (k_l) (Johnson, 1983a). However, such attempts were rather unproductive in the absence of experimental measurements of the changes in body composition or variations in the efficiency value. The former has been remedied by the work of Gibb *et al.* (1992) on changes in the body composition of lactating Holstein–Friesian dairy cows, their results being incorporated in the current advisory manual (AFRC, 1993).

3.2.2   Protein

Alderman (1987) has compared rations for the same production circumstances, calculated using the ARC (1980/84) system, with other alternative published systems from various European countries and North America. In those comparisons, ARC (1980, 1984) yielded estimates of protein requirements (either in terms of dietary crude protein or absorbed protein) which were substantially less than those estimated by any of the other systems. The

Table 3.1. Safety margins for the application of the ARC (1980) metabolizable energy system to dairy herds.[a]

Proportion of cows underfed (%)	Safety margin (%) (i.e. increase in ME allowance)
5	19.7
10	15.4
20	10.1
40	3.0

[a] AFRC (1990).

required amounts of protein calculated using ARC (1980) or ARC (1984) also appeared to be low in relation to observed performances of dairy cows supplied with different amounts of protein (Oldham, 1984; MacRae *et al.*, 1988). In comparison with the DCP system, however, initial appraisals in ARC (1980) show that the system has distinct advantages, especially with regard to the assessment of likely response to non-protein nitrogen (NPN).

Limited collaborative studies have been undertaken with the specific objective of testing both the ARC protein system as a whole and some of its component parts (Oldham *et al.*, 1985; Strickland and Poole, 1985; Thomas *et al.*, 1985; AFRC, 1992). While incremental non-ammonia nitrogen (NAN) supply to changes in dietary protein were predicted with reasonable accuracy, there were substantial errors in the prediction of absolute NAN supply. Also, there were large errors and biases between observed milk protein outputs and those predicted by ARC (1984). As shown in Fig. 3.1 the protein system under-predicted output at low levels of milk protein yield and over-predicted at higher levels. Guidelines on 'safety margins' to cope with these deviations could be provided by assessing individual animal variability in terms of the SD of differences between observed and predicted output. These are not yet available.

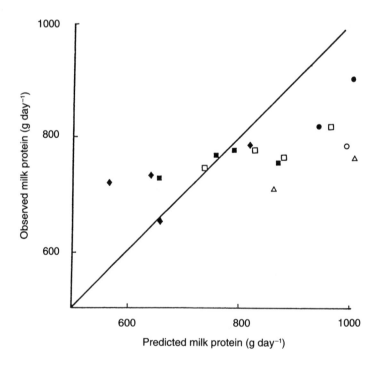

Fig. 3.1. Predicted (ARC, 1984) and observed yields of milk protein. Data from Oldham *et al.* (1985) and Thomas *et al.* (1985).

More recent studies (Metcalf *et al.*, 1996a; Christie *et al.*, 1998) have shown that, for diets which are clearly limiting for protein, the AFRC (1992) system accounts for cow performance very well indeed, and that increments of MP above truly limiting circumstances promote responses in yields of milk constituents, which, in large part can be accounted for in terms of the effects of additional dietary protein and ME supply.

The conceptual basis of the new protein requirement systems is supportable – especially the separate considerations of dietary protein use within the rumen, followed by a consideration of AA use within the body. The systems as currently presented, however, are flawed in detail. The nitrogen needs of rumen microbes vary in form and degree, with a variety of circumstances (see Hobson, 1969). Variable rumen microbial growth yields are widely recognized but have proved difficult to reconcile. In the particular case of dairy cows, Oldham (1984) identified a generally enhanced rumen microbial need for N (in relation to fermentable CHO consumption) as being the origin of the many observations that variation in dietary crude protein intake can have substantial and practically important consequences on diet digestibility and associated food consumption.

Another major flaw in the structural logic of all current systems, is in the description of the efficiency of AA use within the body. MacRae *et al.* (1988) examined a number of experiments in cows and goats where abomasal infusion of casein had been given and from which estimates of efficiency of use can be calculated. From a total of 14 experimental treatments, it appears that only in one situation was the response equal to expectations, based on either ARC (1984) or ARC (1980). In at least eight of the experiments, response in milk yield was less than half of that which would be predicted from the estimates of efficiency (0.85) provided by ARC (1980). Part of this failure lies in variation in the partition of increments of protein (under protein-limiting conditions) such that not all of the response to additional protein is observed in milk. This is seen particularly clearly in the work reported by Whitelaw *et al.* (1986) where response in milk nitrogen was low (in comparison with expectations from ARC (1980, 1984)) but a combination of milk nitrogen plus body nitrogen retention identifies an overall efficiency of utilization which is close to that suggested by ARC (1984). Partition of AAs between alternative routes of synthesis is clearly important in determining the overall efficiency of use for specific processes.

Allowing for problems of the unpredictability of partition, the argument for suggesting constant efficiencies of total AA use in the body under protein-limiting conditions, is clearly not sustainable in the light of well-established principles of AA nutrition which are used routinely in non-ruminant species. The gesture in ARC (1984) towards recognition of two categories of AA (essential and non-essential) as a means to cope with variation in the 'quality' of AAs required, is equally unjustified. It is the provision of the first limiting AA or the extent to which the mixture supplied matches an 'ideal amino acid profile' (ARC, 1981) which requires identification in order that efficiencies of use can become predictable in circumstances where AA supply limits performance. Oldham (1987) identified two elements which contribute to overall efficiency of use and which in combination might be used to estimate effective efficiencies of use in defined circumstances of protein limitation. The first factor is the efficiency with which an ideal AA mixture is used for a particular animal function (maintenance, tissue gain, milk secretion, wool synthesis, etc.). The second factor is the value of the AA supplied relative to that of the ideal AA mixture. An assessment of overall efficiency would therefore need to take into account the balance of processes for which AAs are used.

The extent to which CHO status affects efficiency of protein use, which is well recognized in non-ruminants (Reeds *et al.*, 1987), may be exaggerated in dairy cows because of the low supply of glucose units from the gut in comparison with a very large metabolic demand for glucose. Efficiency of AA use may therefore be sensitive also to the supply of other major nutrients – yet another factor which is not accommodated within the current systems of prediction (ARC, 1980, 1984; AFRC, 1992).

These different issues call for a substantial revision in the framework for accounting for the use of absorbed AAs. Components of that framework will be assessments of limiting efficiencies of use (animal characteristic) and rules to predict the partition of AAs between catabolism and anabolism (as a consequence of AA imbalance), the partition of AAs between competing synthetic pathways and the influence of other nutrients' (especially glucose) availability on AA disposal. The inadequate abilities of current systems always to account for cow performance when diets appear to be limiting in protein may reflect a need to incorporate some of these principles into the accounting procedure. It must be remembered that the systems which exist at present were designed to allow the calculation of protein needs for a defined level of performance. They were not designed to allow the prediction of performance from a defined allowance – that calls for a system of nutrient partition to be included.

3.3   Relevance and applicability of current systems

Dairy cow nutrient requirement systems based on the ARC ME and AFRC MP systems (or NRC (1988) and the CNCPS model; Fox *et al.*, 1992) assume that the characteristics of the lactation cycle are fixed in respect to milk yield, milk composition and body weight, with feed acting as a servant to sustain predetermined levels of performance. Further, it is assumed that each day within the lactation cycle can be regarded as independent. However, these assumptions are challenged by the findings of many feeding experiments. In an early analysis of published experiments, Yates *et al.* (1942) demonstrated that increases in milk yield in response to increased energy supply could be obtained at levels of energy intake considerably above those recommended by feeding standards. This observation, together with those of subsequent studies, showed that cows respond to changes in plane of nutrition in terms of both milk output and body weight change, and that an increase in energy intake produces a negatively curvilinear response in milk yield and a positively curvilinear response in body weight (Jensen *et al.*, 1942; Blaxter, 1950, 1956, 1967; Broster *et al.*, 1979). Concurrent analyses, particularly those of Burt (1957), demonstrated that the milk yield response was a function of the current yield and that the relationship applied equally to cows of different yield potential at similar stages of lactation, and to individual cows between stages of lactation.

Similarly, there is evidence of curvilinear responses in milk output to dietary protein intakes in excess of DCP standards, and that high-yielding cows are more responsive than low-yielding cows (Broster and Bines, 1974). Moreover, Thomas *et al.* (1985) and Oldham *et al.* (1985) have shown increases in yield to increasing dietary protein levels (as MP) above the standards of ARC (1980, 1984). The relationship between protein supply and body weight change is, by comparison, more complex. Most studies indicate that at 'normal' planes of nutrition body weight increases with dietary protein supply (Danfaer *et al.*, 1981). However, in early lactation, with cows deliberately underfed to produce dietary deficiencies in energy and protein, tissue mobilization and body weight loss may be accelerated by increasing the dietary protein supply (Ørskov *et al.*, 1977; Whitelaw *et al.*, 1986).

There is also ample evidence to challenge the concept that milk composition is solely an attribute of the cow. Thus the composition of the diet has been shown to have a major influence on the relative concentrations and outputs of milk fat, protein and lactose. Both Burt (1957) and Broster (1969) observed multi-component responses in the yield of milk constituents to a change in plane of nutrition. Increasing the concentrate allowance led to a greater response in milk protein yield than in milk fat yield and to an increase in the protein content and a reduction in the fat content of milk. These effects of varying the forage:concentrate ratio are schematically represented in Fig. 3.2. However, in many experiments level of energy intake has been confounded with the source of the energy, i.e. with major changes in the chemical composition of the diet. The separate and joint effects of changes in the amount and composition of the diet on the yield of milk fat and non-fat milk constituents and on body weight are illustrated in Fig. 3.3. The results demonstrate that diets high in concentrates tend to favour body gain at the expense of milk fat yield but that body gain and solids-not-fat yield both increase with increasing energy supply, the changes in solids-not-fat reflecting principally an increase in protein yield.

Many other dietary factors have been implicated in modifying the relative output of milk constituents and milk composition. For example, the composition of the concentrate in terms of the source of the starch

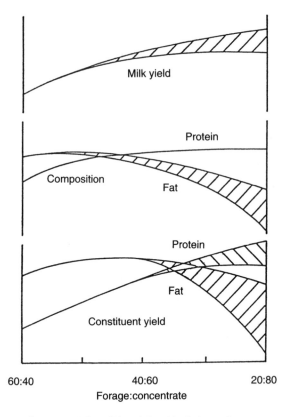

Fig. 3.2. Diagrammatic representation of the relationships between forage:concentrate ratio and milk yield and composition (Sutton, 1986).

(Sutton *et al.*, 1980), the proportions of starch and fibre (Sutton *et al.*, 1987) and the content of both 'free' and 'rumen-protected' lipid (Bines *et al.*, 1978) have all been shown to affect milk composition. Further, the species of forage and its digestibility can also influence milk fat and protein content (Thomas, 1984). The above multi-component effects associated principally with the source of energy can also be demonstrated in response to increases in the supply of dietary protein (see Table 3.2). These data also serve to emphasize the variability in the response in the ratio of fat:protein output. Further, there is evidence to support an interaction between protein supply and source of CHO in the diet in the relative responses in milk constituent output. In this respect Lees *et al.* (1990) observed that in cows in early lactation, increases in milk fat yield in response to increased protein supply were only apparent with rations conducive to high milk fat content (i.e. 40% concentrate diets) whereas increases in protein and lactose output were independent of the type of diet.

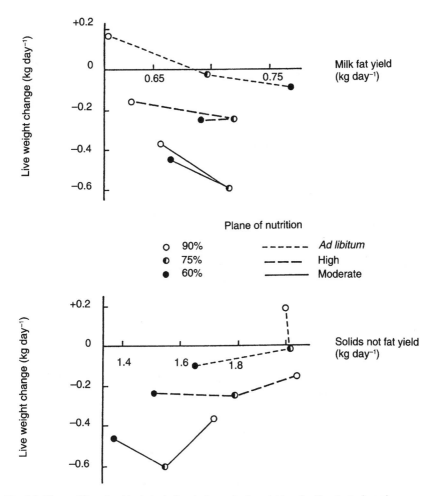

Fig. 3.3. The partition of nutrients by heifers between body weight and milk output when given three amounts of diets containing differing proportions of concentrate to forage (Broster *et al.*, 1979).

The above effects on milk output, milk composition and body weight change resulting from variations in energy and protein supply represent responses obtained over the short term. However, it should be pointed out that a change in input is not immediately reflected in a change in output and that current response may be influenced by previous nutrition. Increases in plane of nutrition require approximately 3 weeks to produce their full effect in milk output and reduction in plane of nutrition up to 6 weeks. These delays are apparent both for changes in the supply of energy (Blaxter and Ruben, 1953) and protein (Thomas *et al.*, 1985). That each day within the lactation cycle is

Table 3.2. A summary of the mean effects on yields of milk fat, protein and lactose of an increased supply of individual end-products of digestion (from Thomas and Chamberlain, 1984).

End-product of digestion	Milk constituent	Effect on yield (response as % of control yield)
Acetate	Fat	+15.6
	Protein	+5.5
	Lactose	+10.3
Propionate	Fat	−10.5
	Protein	+4.0
	Lactose	−0.8
Butyrate	Fat	+10.6
	Protein	−2.9
	Lactose	−7.7
Glucose	Fat	−5.2
	Protein	+4.4
	Lactose	+9.9
Long-chain fatty acids	Fat	+14.9
	Protein	ND
	Lactose	ND
Amino acids	Fat	+4.8
	Protein	+12.8
	Lactose	+15.4

ND, not determined.

not independent is well recognized (Broster and Broster, 1984). However, quantification of the relationship between the short and long term remains a problem. In an analysis of 46 experiments (Broster and Thomas, 1981) the relationship between the immediate and residual effect of a change in feeding depended on the basal plane of nutrition in the 'immediate' period. For low and medium planes of nutrition the residual effect amounted to 0.55 of the immediate response. For cows initially offered high planes of nutrition no relationship existed between short- and long-term responses, but seven of the experiments demonstrated negative effects on subsequent output indicating compensation in mid and late lactation. Thomas *et al.* (1985) could not demonstrate any residual effect arising from the manipulation of dietary protein supply in early lactation. Some of the experiments also provided evidence of residual effects on milk composition, but only in terms of solids-not-fat (principally in protein); there were apparently no residual effects on milk fat concentration. In general, when residual effects were detected they appeared to be associated with a long-term change in the partition of nutrient use by the cow. Thus the additional milk output during the residual period was associated with a reduction in body weight gain.

The results from the feeding experiments which have been examined in this section challenge the concepts underlying the application of current feeding systems by demonstrating that:

1. Milk production and body tissue synthesis/mobilization are dynamic and inter-related processes, and that one cannot be pre-determined or fixed to allow the requirement for energy and protein to be predicted for the other.
2. The relative output of individual milk constituents and the composition of milk are influenced by the amount and composition of the diet and thus are not solely an attribute of the cow and her physiological state. Further, the effects of diet on the outputs of individual milk constituents cannot be related to feed composition expressed simply in terms of ME and rumen-degradable and undegradable protein contents.
3. Each day within the lactation cycle cannot be considered to be independent. The partition of nutrients between milk and body tissue depends not only on current nutrition but also on the cow's previous nutritional history as reflected in her current state.

It is important to recognize that these limitations do not question the tenets of the ARC models of energy and protein requirements. Thus the ME system is sufficient for the purpose of predicting total energy output, at least in the long term (AFRC, 1990). Less information is available adequately to test the protein system and in any case its objectives are to define minimum requirements for MP to achieve a given output of milk protein. The argument presented here is not that the systems are inaccurate but that that which they seek to predict lacks relevance to the present and future needs of milk producers.

4. Towards a New Approach

4.1 Purpose of feeding systems

Feeding systems are used for two main purposes: to provide an objective basis for assessing the nutritive value of feeds; and to provide a mathematical basis for relating the amount and composition of the diet to the cow's current or future productive performance. With regard to the latter, all existing feeding systems are principally directed towards the question: 'How much dietary energy and protein does the cow require to meet her needs for maintenance and a given energy and protein output in milk?'. In practice, the crucial question is generally a different one, since there is no obligation for the farmer to meet the cow's theoretical requirements if this is contrary to economic profitability. Rather, the need is to know the amounts of the available feeds which constitute a diet offering (over the long term) the maximum differential between feed costs and returns on milk sales. This necessarily implies an economic analysis of the cow's responses in milk yield, composition, body-weight change and fertility to changes in the composition of the diet and level of feeding, recognizing both short- and long-term effects.

Under modern conditions most dairy cows are either given rationed allowances of concentrate feeds and forage *ad libitum*, or *ad libitum* access to a complete mixed diet. Prediction of feed intake, and of forage intake in particular, must therefore be seen as an integral part of any system designed for response analysis. Also, since the partition of dietary energy and protein between milk secretion and body tissue deposition is determined by both dietary and animal factors, these must be taken into account in the prediction of milk production and live-weight gain. Finally, since the payments for milk to most producers in the UK are now based on pricing schemes with differing and variable values for milk fat, protein and lactose, predictions of milk production would best be expressed in terms of yields of individual milk constituents rather than in terms of volume or weight of milk, as with present feeding systems.

4.2   Specifications for a feeding system

In order to be relevant to the needs of the dairy farmer, a dairy cow feeding system should predict short- and long-term effects on:

- Voluntary feed intake.
- Partition of nutrient use between milk secretion and body tissue synthesis.
- Yields of milk fat, protein and lactose.

Additionally, the system should recognize circumstances where the shortfall in nutrient supply relative to the cow's biological drive for milk production is likely to lead to a loss in homeostasis and the occurrence of metabolic disorders. Furthermore, in view of the increasing emphasis on the relationship between diet and health in man and the large proportion of milk that is used for manufactured foods, it would be useful if the system had the structural flexibility to incorporate predictions of further important aspects of compositional quality, such as fatty acid composition, or specific milk proteins for cheese making.

A starting point for the development of such a system is the recognition that the cow's productive performance depends on the supply of nutrients derived from the diet and utilized for milk secretion and body tissue synthesis. However, both nutrient supply and nutrient utilization are modulated by the characteristics of the cow (in particular her physiological state, which is influenced by both genetic and environmental factors). Against this background the following sections consider nutrient supply, nutrient utilization and cow characteristics with a view to their quantitative assessment and description in terms that could be incorporated into a feeding system. There is no corresponding consideration of feed intake since this has been the subject of a separate TCORN review (AFRC, 1991). However, it should be stressed that there is growing evidence that changes in feed intake are an integral component of the cow's complex physiological reaction to a change in diet and that it is biologically incorrect to isolate feed intake from the other functional responses that are considered here.

It is the view of the Working Party that a satisfactory representation or framework for prediction of milk production responses in dairy cows will need to be dynamic (i.e. capable of predicting events over time) whilst being as simple as possible to meet the stated objectives. The most appropriate level of representation is a subject for debate.

Any new framework for response prediction needs to have one major element representing nutrient supply and a second representing nutrient utilization, with transactions occurring in the context of a description of the cow's current state. A framework of this type is shown schematically in Fig. 4.1 and the biological and mathematical description of its components are considered in the remainder of the report.

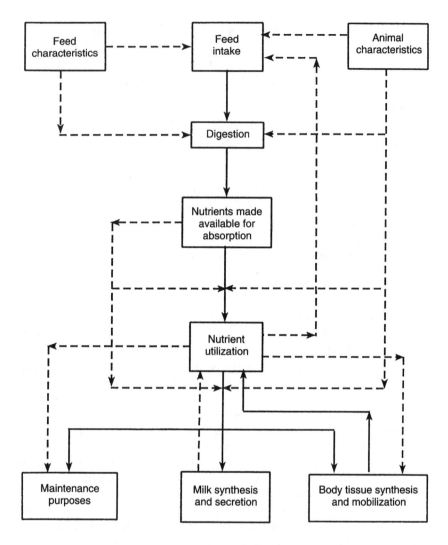

Fig. 4.1. A schematic representation of the effects of feed and animal characteristics on feed intake and utilization of the dairy cow. → flow of nutrients; – – → regulatory influences operating at various points.

5. Nutrient Supply

5.1 Digestion in the ruminant

The relationship between the amount and composition of the diet and the supply of nutrients absorbed from the gastrointestinal tract in ruminant animals is complicated by the nature of the digestive processes and particularly by the activities of the symbiotic populations of microorganisms found principally in the reticulo-rumen (rumen) but also in the caecum and colon.

Microbial activity in the rumen results in degradation of dietary carbohydrates with formation of volatile fatty acids (VFA) and methane, degradation of dietary protein with formation of ammonia (NH_3) and amino acids, hydrolysis and bio-hydrogenation of dietary lipids, and synthesis of microbial biomass. VFA are absorbed largely from the rumen and methane is lost by eructation, whereas undigested feed residues and microbial biomass pass onwards for subsequent digestion and absorption in the small intestine. Materials passing from the small intestine undigested may undergo further fermentation in the caecum and colon, with some further absorption of VFA and ammonia and an associated loss of microbes and undigested feed and residues in the faeces.

Metabolic events within the rumen are central to the supply of nutrients to the host animal, affecting not only those nutrients absorbed from that organ but also the amount and composition of the undigested feed and microbial materials which pass onwards for subsequent digestion in the post-ruminal tract. The rumen can be considered as a discontinuous microbial fermenter in which the breakdown or modification of dietary carbohydrate, protein and lipid components, the formation of fermentation end-products and the production of microbial biomass represent a series of highly interrelated processes. These are governed by the nature of the feed substrates consumed and by the composition and metabolic activity of the microbial population, which itself is greatly influenced by the chemical composition and physical form of the diet, and by the frequency of feeding (Sutton, 1985).

The last two decades have brought major advances in techniques to quantify nutrient supply and, whilst the limitations of accuracy of some of the techniques must be recognized, their application has allowed a framework of understanding of quantitative aspects of digestion in the rumen, small intestine and caecum and colon to be established. Nutrient supply has been assessed using three main approaches:

1. Measurement of the disappearance or apparent absorption of feed and microbial constituents in different sections of the digestive tract.
2. Measurement of VFA production.
3. Measurement of the entry of nutrients into the portal venous system.

The first approach has been used mainly to determine firstly the proportion of digestion which occurs in the stomach, small intestine and caecum and colon, and secondly the apparent absorption of specific nutrients (AAs, fatty acids and glucose) from the small intestine. The procedures depend on estimation of the amounts of specific constituents consumed in the feed, passing from the stomach to the small intestine, passing from the small intestine to the caecum and colon and excreted in the faeces. Technically the procedures involve the use of animals surgically modified with cannulae of varying designs fitted in the abomasum or duodenum and ileum (Faichney, 1975; MacRae, 1975; Komarek, 1981). Calculations of digesta flows depend on the type of cannula used. With simple T-shaped cannulae where the collection of a representative sample of digesta passing the cannula can be problematical, the flow of nutrients is best calculated by reference to two indigestible unabsorbed markers associating themselves with the liquid and the particulate phases of the digesta, respectively (using the dual phase marker technique; Faichney, 1975). Where cannulae are of the type which can force all digesta to exit the body, less rigorous marker techniques can be applied (MacRae, 1975), and indeed automated recording and sampling machines can be used, thus eliminating the need for marker corrections (Zinn *et al.*, 1980). Using these techniques, 'disappearances' of AAs, fatty acids and glucose (starch) from digesta between the duodenum and ileum have provided a reasonably precise measure of the net absorption of these constituents by the animal, although questions remain about the relationship between 'net (apparent) absorption' and 'true absorption' values because of the problems of quantifying the non-resorbed endogenous secretion component of ileal digesta.

In contrast, the 'disappearance' of constituents in the rumen and caecum and colon provides only a crude indication of the quantities of total VFA that are formed at those sites, and alternative methods are required to provide information on the production rates of individual VFA. Several methods have been developed, including direct methods based on *in vivo* isotope dilution measurements using ^{14}C-labelled VFA, and indirect methods making use of the stoichiometric relationships which exist between the production of VFA and the production of methane (Webster, 1978) or between the production of

VFA and the amount and composition of the mixture of substrates fermented. In this regard Murphy *et al.* (1982) proposed a series of preferred equations for relating the composition of the substrates fermented to the proportions of VFA observed in the rumen (Table 5.1). However, as has been discussed by Sutton (1985), the proportions of VFA in the rumen may not precisely reflect their relative production rates, especially when the diet contains a substantial proportion of concentrates.

An alternative approach to estimating the disappearance of nutrients from or the production rate of nutrients in the gut is to determine the uptake of nutrients into the venous drainage of the gut, i.e. at the portal vein, using a simple arterio-venous (A–V) difference technique. This involves two types of measurement. The A–V difference in concentration of a nutrient in the blood supplying (arterial) and draining (portal) the gut (generally the aortal or carotid arterial blood and the portal venous blood) must be determined, and simultaneously there must be a measurement of blood flow in the portal vein. Generally speaking, most blood metabolite concentrations can be measured routinely with high accuracy, but techniques to measure blood flow are less reliable and there is a need for further development of the methods. In theory, blood flow can be measured by several techniques, including downstream dilution of markers such as para-aminohippuric acid (PAH; Roe *et al.*, 1966) or indocyanate green (Wangsness and McGilliard, 1972), downstream thermo-dilution (White *et al.*, 1967), or the electromagnetic or Doppler electronic flow probes (Prewitt *et al.*, 1975; Durand et al., 1988; Huntington *et al.*, 1990). Unfortunately, in practice, it has proved difficult to establish good flow probes on the portal vein of cattle. Huntington *et al.* (1990) reported a

Table 5.1. Stoichiometric relationships between substrate fermented and VFA formed (Murphy *et al.*, 1982).

Equation no.	Substrate	Mol of product formed per mol of substrate				
		Acetate	Propionate	Butyrate	Branched	Valerate
1	Soluble carbohydrate[a]	1.38	0.41	0.10	—	—
2	Soluble carbohydrate[b]	0.90	0.42	0.30	0.10	—
3	Starch	0.80	0.60	0.20	—	0.10
4	Hemicellulose	1.58	0.12	0.06	—	0.09
5	Cellulose	1.12	0.51	0.11	—	0.07
6	Pectins	1.38	0.41	0.10	—	—
7	Amino acids[a]	0.40	0.13	0.08	—	0.33
8	Amino acids[b]	0.36	0.16	0.08	—	0.33

[a] Fermentation in high forage diets.
[b] Fermentation in high concentrate diets.

45% reduction in blood flow with probes (i.e. versus PAH dilution) in steers, due to anatomical constraints which precluded proper placement and functioning of the flow probe. Presently, therefore, dye dilution is the technique most commonly used to obtain frequent and reproducible measures of portal blood flow (see review by Seal and Reynolds, 1993).

5.2   Diet and nutrient supply

Application mainly of the first two techniques described above has provided valuable information on the disappearance of dietary constituents in the different regions of the alimentary tract and on the products of digestion formed. This information clearly allows a framework of principles to be established to describe the digestive breakdown of dietary carbohydrates, lipids and proteins in the rumen, small intestine and caecum and colon. However, the corresponding framework of principles to describe VFA production is subject to more uncertainty. There are also difficulties in establishing simple guidelines for the dietary factors influencing the supply of AAs and LCFA to the small intestine since both depend heavily on the influences of diet on microbial growth in the rumen, and those influences are both subtle and complex.

There is now a plethora of ruminant protein requirement systems developed throughout Europe and in the United States (e.g. the UK (ARC, 1980, 1984; AFRC, 1992); France (INRA, 1988); Scandinavia (Madsen, 1985); USA (NRC, 1985; Sniffen *et al.*, 1992) and Holland (Taminga *et al.*, 1994)) all of which recognize that the passage of AA-N to the duodenum may be greater or less than the dietary N intake, the balance between the two depending on the amounts and rumen degradability of N in the diet relative to the fermentable energy sources that are available to the rumen microbes. However, within the context of developing predictive relationships, the nature and digestive fate of the individual fermentable components in the diet and the influences of diet composition on the growth characteristics and fermentation activities of the rumen microbial population need also to be considered.

Similarly, the amounts of LCFA passing to the duodenum in cows are generally greater than the amounts consumed in the diet, reflecting the contribution of rumen microbial lipid synthesis. However, the extent of that synthesis is variable and is influenced not only by the amount and type of lipid in the diet (Sutton *et al.*, 1983) but also by the source of starchy carbohydrate (Brumby *et al.*, 1979) and, for some diets, by the rumen availability of methionine (Chamberlain and Thomas, 1982). Moreover, because of their physico-chemical influences on rumen fermentation, dietary inclusions of lipids may impair microbial breakdown of fibrous carbohydrates and reduce both microbial protein synthesis and VFA production (McAllan *et al.*, 1983). Unsaturated fat supplements also modify the pattern of VFA produced, increasing the proportion of propionate formed (Thomas and Chamberlain, 1984).

From the above it is evidently not feasible to summarize the relationship between diet and products of digestion in a short succinct statement or a few simple mathematical expressions. However, as background to later discussions it would be helpful to outline some of the main features of the relationship, which are summarized as follows:

1. Typically, 0.55–0.70 of the digestible energy (DE) in the diet 'disappears' in the rumen, 0.20–0.35 in the small intestine and 0.05–0.15 in the caecum and colon. This is influenced by the chemical composition of the diet, its physical form and the level of feeding. For finely ground forage diets the proportion of digestion in the rumen may be reduced and the proportion of digestion in the caecum and colon increased outside the ranges indicated.
2. Approximately 0.75–0.88 of the energy 'disappearing' in the rumen and caecum and colon is converted to VFA, the mixture of VFA formed depending on the diet and the microbial population which develops. Variation of fermentation in the rumen is much greater than in the caecum and colon.
3. With diets solely of forage the molar proportions of VFA formed in the rumen are typically 0.65–0.75 acetate, 0.15–0.25 propionate and 0.08–0.15 butyrate. Inclusion of cereals or sugar-rich feeds generally reduces the proportion of acetate, with compensatory increases in propionate, or less frequently, propionate and butyrate. Butyrate proportions rarely exceed 0.30, except on diets containing very readily fermented starch sources, where propionate levels may approach 0.45.
4. The ruminal fermentation of soluble sugars and pectins is virtually complete, and with most forms of starch less than 0.1 of that ingested passes unfermented to the small intestine. Ground maize diets are an exception and as much as 0.35 starch may escape fermentation. Starch passing to the small intestine is almost totally digested and absorbed as glucose within the small intestine, except in cattle given whole grains where there may be caecal fermentation and even starch loss in the faeces. With increased quantities of starch passing to the small intestine, it is possible to exceed the capacity of the small intestine to digest starch.
5. With diets composed solely of long or chopped forages 0.85–0.95 of digestible cellulose and hemicellulose is fermented in the rumen and the remainder in the caecum and colon. When forages are given in ground pelleted forms or where the forage is supplemented with starchy feeds, ruminal fermentation of fibrous carbohydrates is reduced and caecal fermentation is increased, often with some overall reduction in digestibility of fibre.
6. AA passage to and absorption from the small intestine depends on the amount of protein consumed, the extent of ruminal degradation of that protein and the synthesis of microbial protein. With low-protein diets, protein flow to the small intestine may exceed protein intake, but with high-protein diets

there may be substantial losses of protein between the diet and the duodenum. The latter may be particularly pronounced with diets of fresh or ensiled forages and with diets containing significant amounts of NPN. With these feeds the high degradability of ingested protein, and a possible limitation of readily fermented carbohydrate sources leads to elevated concentrations of ammonia in the rumen and substantial ammonia absorption (MacRae and Beever, 1997). Similar losses may occur with high-grain diets when adverse conditions in the rumen reduce the efficiency of microbial capture of nitrogen.

7. Supplementary protein feeds of inherently low rumen degradability have direct effects on the amount of undegraded dietary protein passing to the duodenum but can also increase protein flow by stimulating carbohydrate digestion and microbial protein synthesis (Dawson *et al.*, 1988). Thus the responses in duodenal protein supply to dietary supplements of low-degradability proteins are not always linear (Beever *et al.*, 1990).

8. The LCFA that are absorbed from the small intestine are derived from dietary sources and from the rumen microbes. The dietary component directly reflects the amounts of fatty acids consumed, since rumen metabolism involves hydrogenation of unsaturated fatty acids but essentially no fatty acid catabolism. In contrast the lipid content of microorganisms can be highly variable and influenced by a range of factors which affect microbial growth, including dietary carbohydrate and protein as well as lipid supply.

5.3   Prediction of nutrient supply in the dairy cow

Whilst current knowledge of digestion in the ruminant provides a sound conceptual framework with which to consider the prediction of nutrient supply, the information available is subject to a number of important limitations. In virtually all published experiments on the subject there is insufficient information about the chemical composition and nutritional characteristics of the diets for the results obtained to be readily summarized and used as a basis for generalization. Furthermore, most of the information available has been derived from experiments with sheep at low levels of feeding, and there remains a need for more quantitative information on the lactating dairy cow receiving diets at three to four times the maintenance level. This is particularly so with respect to protein digestion, where much of the data available refer to the passage of total N or NAN to the duodenum rather than to individual AAs. There is a widely held view that the AA composition of duodenal digesta in the ruminant varies little because of the large contribution derived from microbial protein. However, those AAs whose content in duodenal digesta is most subject to variation, particularly methionine and lysine, often appear to be most limiting for milk production in the cow, and their supplies could have important modulating influences on animal performance.

In general, it can be concluded that the effects of diet on nutrient supply in the cow could be quantified using current methodology, although significant technical problems still surround measurements of the production of individual VFA in the rumen (Sutton, 1985). Whether the need for these measurements can be met through the use of a stoichiometric approach similar to that of Murphy *et al.* (1982) warrants careful examination. The paucity of data to define relationships between diet and nutrient supply in dairy cows is of concern in respect of the development of satisfactory nutrient-based feeding systems for dairy cows.

6. Nutrient Use

6.1 General aspects of ruminant metabolism

The composition of nutrients absorbed from the digestive tract of ruminants differs markedly from that in non-ruminants. A major distinguishing feature in the ruminant is that the direct contribution from carbohydrate is generally less than 0.10 of total DE and LCFA are also a rather low overall proportion of DE (0.05–0.10) in comparison with non-ruminants. With AAs generally supplying 0.15–0.25 of DE, the majority of DE is in the form of VFA absorbed from the fermentation areas of the gut. Of these, acetate (0.25–0.35), propionate (0.15–0.30) and butyrate (0.08–0.15), comprise the major part of the digested energy.

The balance of these major nutrients has been shown to have important influences on the partition of nutrient use between the mammary gland and body tissues (see for example Coppock *et al.*, 1964; Ørskov *et al.*, 1969, 1977; MacRae *et al.*, 1988) as well as on the relative yields of fat, protein and lactose in milk (Tyrrell, 1980; Thomas and Chamberlain, 1984). The relative proportions of acetic, propionic and butyric acids supplied as fermentation products seem to be the single most important feature of nutrient balance which influences nutrient partition (Sutton, 1986). Recognition of VFA profile as a central axis, through which the chemical and physical composition of the feed and feeding strategy might alter nutrient balance and thence nutrient partition, might present an opportunity to modify current feed descriptions in such a way as to enable the prediction of partition by linking predictive relationships between food composition/feeding system, rumen VFA proportions and nutrient partition. There may therefore be a practical possibility that prediction of the yield of milk constituents from existing feeding systems could be improved through modifications designed to take account of the products of digestion absorbed by cows receiving diets of defined chemical composition and the system of feeding, this being a major advance on ME as a standard and unique entity.

Alternatively, it may be necessary to develop systems of response prediction that deal with nutrient use in terms of the actual amounts (masses) of the individual nutrients absorbed. This will be because of the distinctive quantitative influences of the supplies of individual nutrients on specific metabolic processes. Such an approach would involve quantitative representations of the physiology and biochemistry of nutrient use in the animal's tissues which hitherto have not generally been regarded as being of central relevance to feeding systems. An overview of those issues which might need quantitative representation follows.

6.2 Nutrient supply and use in body tissues

End-products of digestion which are removed from the lumen of the gut are presented for metabolism in the gut tissues, the liver, peripheral tissues and the mammary gland. As the tissues of the gut contribute 0.18–0.28 of oxygen consumption (i.e. thermogenesis) (Reynolds, 1995), the amounts and proportions of nutrients appearing as substrates arriving at the liver will differ from that absorbed from the lumen of the gut. This may be particularly true for glucose, the gut being an extensive site of glucose metabolism, and for certain AAs, especially glutamine and glutamate (Bergman, 1986; Reynolds *et al.*, 1994). While medium- and long-chain fatty acids (C10–C12) may not be subject to substantial metabolism within gut tissue their route of entry into the arterial circulation via the thoracic duct avoids initial metabolism via the liver.

6.2.1 Basal nutrient use

First demands on nutrient supplies will be those processes which can be identified as obligatory, of which basal metabolism and certain aspects of fetal development might be considered parts. Basal requirements for 'energy' for thermoregulatory processes, maintenance of muscle tone (including cardiac function), etc., may be fulfilled through the production of ATP from oxidation of any of the major nutrients and, as such, are not nutrient specific. Certain other functions, for example, some aspects of central nervous system activity, are more nutrient specific, especially as relates to glucose provision, and hence an apparent obligatory need for glucose to meet 'basal requirements'. Depending on the assumption made, the requirement for glucose for basal metabolism seems likely to be in the region of 2 g kg^{-1} $W^{0.75}$ day^{-1}. There is also a basal endogenous loss of nutrients from the animal (ARC, 1980, 1984; AFRC, 1992) which represents obligatory demand on nitrogenous substrates; it is usually assumed that this obligatory loss is met from AAs.

6.2.2   Nutrient use for fetal growth

The fetus and placental tissue take high priority in the demands on maternal nutrient supply; while these are small in early gestation, they become substantial as term approaches. The fetus uses a restricted range of substrates. Girard and Ferre (1982) found no detectable uptake of free fatty acids or 3-hydroxybutyrate by the fetus near term but substantial use of glucose, lactate (formed from glucose in the placenta), AAs and acetate. Where nutrient supply is low, AAs appear to become the predominant metabolic fuel for the fetus – an issue of nutritional consequence when fetal burden is high. The fetus represents only a fraction of the gravid uterus as a whole and relatively few quantitative data are available on placental nutrient use.

6.2.3   Nutrient use for milk production

The synthesis of milk fat, protein and lactose from substances extracted from the arterial blood supply takes place in the alveolar cells of the mammary gland. The main pathways of synthesis are outlined schematically in Fig. 6.1 and detailed descriptions of the major pathways are given in Mepham (1982), Rook and Thomas (1983) and Mepham (1987).

A summary of the quantities of individual nutrients taken up by the mammary gland is presented in Table 6.1, based on cows yielding between 16 and 20 l milk day^{-1} (Bickerstaffe *et al.*, 1974).

Of the fatty acids secreted in milk fat, all of the short-chain and part of the medium-chain fatty acids (up to C16) are synthesized *de novo* in the udder from acetate and 3-hydroxybutyrate, the latter contributing both via cleavage to acetate and by incorporation as an intact C4 unit (Moore and Christie, 1979). The remaining medium-chain and all the long-chain fatty acids are absorbed by the udder pre-formed from the blood plasma low-density lipoprotein triglyceride and free fatty acid fractions (Moore and Christie, 1979; Annison, 1983; Dils, 1983). These pathways distinguish ruminants and non-ruminants in that glucose makes little contribution as a precursor for fatty acid synthesis because of biochemical limitations on metabolism within the gland (Smith and Taylor, 1977; Smith *et al.*, 1983). Nutrient provision can alter fatty acid metabolism in the gland through alterations in free fatty acid levels (see for example Annison *et al.*, 1974; Bickerstaffe *et al.*, 1974) but direct dietary effects are usually obscured by the superimposition of fatty acids of immediate tissue origin ('mobilized fat') whose contribution may be expected to vary inversely with the overall input of fatty acids on their precursors (acetate and butyrate) from the diet.

AAs in arterial blood are the principal precursors of proteins in milk (Mepham, 1982) although small peptides may also make a contribution (Backwell *et al.*, 1994). Some AAs (methionine, tryptophan, phenylalanine) are considered to be transferred to milk almost quantitatively (Metcalf *et al.*,

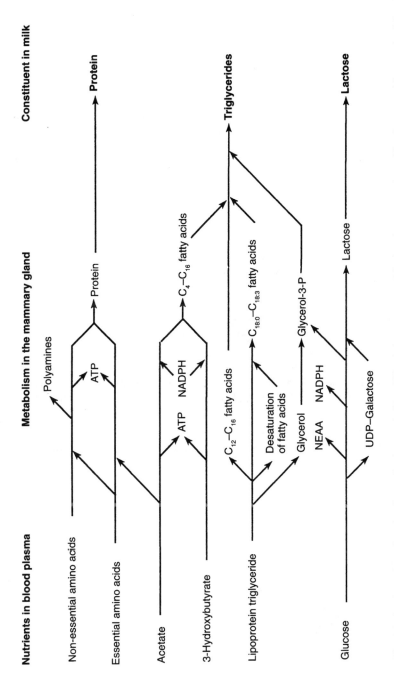

Fig. 6.1. A schematic outline of pathways for the metabolism of blood plasma acetate, 3-hydroxybutyrate, amino acids, glucose and lipoprotein triglycerides in the mammary gland and for the synthesis of milk fat, protein and lactose.

Table 6.1. A summary of the main substrates taken up by the mammary gland of the dairy cow and used for the secretion of milk fat, protein and lactose (from Bickerstaffe *et al.*, 1974).

Substrate	Mammary gland uptake (g day^{-1})
Glucose	1243
Acetate	398
3-Hydroxybutyrate	218
Triglycerides	290
Amino acids	457

Notes: Data (except amino acids) are mean values based on 19 observations in six cows, mean milk yield 19.5 kg day^{-1}. Milk fat, protein and lactose yields were 698, 634 and 938 g day^{-1}, respectively.
Values for amino acids are based on 16 observations with four cows, mean milk yield 16.9 kg day^{-1} and milk protein yield was 386 g day^{-1}.

1996b) whilst some are catabolized (especially the branched chain AAs) and yet others synthesized within the gland. There is no consistent evidence to suggest that arterial concentration of AAs is particularly sensitive to dietary manipulation, nor that blood AA concentrations relate closely to milk protein synthesis (Bequette *et al.*, 1994; Metcalf *et al.*, 1994, 1996a).

Glucose is the main precursor of milk lactose (Mepham, 1982). Glucose also serves as a substantial metabolic fuel within the gland and may account for about 0.25 of total CO_2 production in the udder (Annison *et al.*, 1974; Bickerstaffe *et al.*, 1974). Under similar circumstances acetate accounts for 0.25–0.33 of udder CO_2 production. Milk volume (and therefore lactose yield) correlates quite reasonably with blood flow (Linzell, 1974) and glucose irreversible loss, but the driving force in such a correlation is not immediately clear. In high-yielding cows it can be estimated that glucose metabolism in the udder may approach 4 kg day^{-1}.

6.2.4   Nutrient use for tissue synthesis

The pathways for the use of AAs in protein synthesis are the same in ruminant tissues as in other species. For some amino acids there may be differences in emphasis in some detailed aspects of metabolism (Buttery, 1977) but these do not appear to be sufficient to lead to an expectation that overall AA use would differ in principle (nor indeed in magnitude) in ruminants as opposed to non-ruminants. The rates of protein synthesis in the whole body and in major tissues appear to be similar in ruminants to that in non-ruminant species when scaled to a metabolic body size (Reeds and Fiorotto, 1990). As much as 0.25

of total body protein in cows (20–25 kg) may be readily labile according to Botts *et al.* (1979) and Bauman and Elliot (1983), which, in a cow with an estimated total body protein mass of 80–100 kg, could be a substantial nutritional 'buffer' in times of nutritional shortage. However, the estimates of Chilliard *et al.* (1987) were appreciably lower, based on a review of published and unpublished data on dairy cow protein mass obtained by a variety of techniques. They suggested a maximum of 15 kg mobilizable protein in early lactation, but that to avoid adverse effects upon milk production, this should be limited to 5–10 kg protein or 0.06–0.12 of the total body protein mass. Their lower estimate was confirmed by Gibb *et al.* (1992), who recorded 0.066 (5.6 kg) of 85 kg total body protein as having been mobilized over the first 8 weeks of lactation, although this does not take account of significant relocation of protein to the gut and liver, undergoing extensive hypertrophy at that time. Andrews *et al.* (1994), using Holstein cows serially slaughtered at pre-partum, early lactation and late lactation, found little change in body protein weight, but estimate that on average 0.07 of body energy mobilized came from body protein, or 3.5 kg over the 77 days of lactation, at which time maximum body energy loss had been recorded.

In ruminants, acetate is the main precursor of *de novo* fat synthesis, whilst LCFA are important substrates under some circumstances. The subcutaneous and inter-abdominal tissues have an essentially similar pattern of substrate use for fat synthesis, although there are some differences between the tissues in the activity of common metabolic pathways (Vernon, 1981). Inter-abdominal fat is more saturated than subcutaneous fat and appears to be metabolically more responsive to alterations in the animal's diet or physiological state. Plasma lipoprotein triglycerides are the main sources of fatty acid incorporation into adipose tissue; LCFA so incorporated may be extended in chain length through an elongation pathway or subjected to a desaturase system with a high affinity for stearic acid. *De novo* fatty acid synthesis may account for about one-third of the total fatty acids deposited in adipose tissue.

Tissue fat may be gained or lost depending on physiological state (especially stage of lactation) and current nutrition, as indicated by Walsh *et al.* (1992), who examined *in vitro* rates of lipogenesis and lipolysis in adipose tissue of cows between calving and week 29 of lactation. As the maximum rate of net lipolysis may be of the order of 3–4 kg fat per day in a 'normal' dairy cow (Konig *et al.*, 1979) the potential for tissue fat to 'buffer' against nutritional adversity by providing non-specific metabolic fuel is clearly substantial. As total body fat may be 100–150 kg (Butler-Hogg *et al.*, 1985; Gibb *et al.*, 1992) this nutritional 'buffer' can be effective for a considerable time, even when functioning at maximum rate, and may explain why in high yielding cows positive energy balance may not occur until mid-lactation (Beever *et al.*, 1998).

6.2.5   Nutrient use for metabolism

Net synthetic pathways carry with them associated metabolic activity which might be referred to as 'enhanced maintenance'. Thus, the rate of protein synthesis has been found to correlate reasonably closely with the rate of ion transport into synthetic cells (Milligan and Summers, 1986). It is debatable whether such associative pathways should be considered as a charge on maintenance (in which case, maintenance becomes variable rather than strictly related to basal metabolism) or a charge on 'performance' (or accretion). In either instance, nutritional accounting and the assessment of response would need a clear view of the nature of the associated pathways, and their particular nutrient demands, in order to derive systems for predicting milk secretion on the basis of nutrient disposal.

Important 'associated pathways' are those of energy (as ATP) needs in relation to synthetic pathways, and the use of secondary substrates, for example, glucose as a source of reducing equivalents and glycerol in triglyceride synthesis from acetate.

6.3   Regulation of nutrient use

A central feature of the metabolism of the lactating dairy cow is that the synthesis of milk constituents and the synthesis or mobilization of tissue fat and possibly tissue protein occur simultaneously and are inter-related.

As defined by Bauman and Currie (1980), homeorhetic controls orchestrate the development of different physiological states while homeostatic controls maintain metabolic equilibrium within the physiological state. The endocrine agents involved in homeorhetic (strategic) and homeostatic (tactical) controls can be the same. In regulating nutrient disposal, the insulin and the growth hormone axis appear to be particularly important (Hart, 1983; Bauman and McCutcheon, 1984; Fullerton *et al.*, 1989) and have both homeostatic and homeorhetic functions.

There are strong indications that variations in rumen VFA proportions, and perhaps supplies of propionate in particular, can influence insulin release (Sutton *et al.*, 1980, 1987; MacRae *et al.*, 1988; Sano *et al.*, 1995). Relative affinities of adipose tissue and the mammary gland for insulin (which change with stage of lactation) thereby come into play in determining the metabolic fate of fat precursors and this may be a main mechanism by which shifts in the balance of major end-products of carbohydrate digestion influence milk fat yield and content and the rate of adipose tissue accretion. Thus, for example, the influences of supplies of specific nutrients on yields of milk constituents (Table 6.2) and the corresponding effects on milk constituent synthesis which are observed in response to the intra-ruminal infusion of propionic acid (Ørskov *et al.*, 1969). The particular controls which regulate disposal of AA

Table 6.2. Yields of milk and solids corrected milk (SCM), yields and content of milk fat, solids-not-fat (SNF), protein and milk energy content (from Ørskov *et al.*, 1969). (Mean values are given for three cows receiving a pelleted basal ration, or having 15.4% of metabolizable energy replaced with acetic or propionic acid.)

Treatment	Milk yield (kg day⁻¹)	SCM (kg day⁻¹)	Milk fat (g kg⁻¹)	Milk fat (g day⁻¹)	SNF (g day⁻¹)	SNF (kg day⁻¹)	Protein (g kg⁻¹)	Protein (g day⁻¹)	Energy (kJ g⁻¹)
Basal	19.19	13.91	19.6	376	86.4	1.66	31.9	612	2.35
Basal + acetic acid	15.58	12.18	25.8	402	85.3	1.33	31.5	491	2.58
Basal + propionic acid	14.65	10.13	19.2	281	83.5	1.22	30.5	447	2.31

supplies between milk protein synthesis, tissue protein accretion and net catabolism are not so well documented. As in growth, there can be consequential effects on tissue fat metabolism as a result of enhancing protein secretion in milk (Ørskov *et al.*, 1977; Whitelaw *et al.*, 1986). Changes in protein supply may attenuate the endocrine system to facilitate the response to provision of necessary nutrients. Thus, MacRae *et al.* (1991) found that insulin-like growth factor 1 concentrations in the blood and the anabolic response of the animal were enhanced to a standard growth hormone (somatotropin) challenge in sheep offered high- as opposed to low-protein diets. Such controls may be pertinent to the substantial shift in amino acid partition between tissue and mammary gland use which appear to be possible within lactation under protein-limiting circumstances (MacRae *et al.*, 1988).

Regulatory mechanisms which respond to current nutrition may be expected to interact with homeorhetic control such that response to nutrition at one stage of lactation may differ from, apparently, the same circumstances later in lactation (e.g. Broster *et al.*, 1985). Representation of such regulation within quantitative nutritional systems must therefore recognize these two forms of regulatory mechanism.

6.4 Prediction of nutrient utilization

Qualitative descriptions of nutrient use are available in abundance. In contrast, quantitative descriptions of major determinants of nutrient use are limited – in part, due to the inadequacy of current techniques (such as A–V difference measurements) and sometimes by lack of relevant observations. There are, nonetheless, clear indications that the balance of major energy-yielding nutrients (especially the relative proportions of fermentation end-products) is a central issue determining the partition of those nutrients between alternative pathways of use. The interplay between homeorhetic/strategic controls on lactational progress and homeostatic/tactical control on current metabolic state is also recognized qualitatively as being a potentially important determinant of nutrient partition; whilst such interactions have yet to be quantified, they will be necessary for predictive purposes. It would seem most likely that such quantification would be in the form of a representation of such events, rather than a precise description. Such a representation would need to recognize the balance of anabolic and catabolic controllers and the impact of nutrient balances on the attenuation of the system. An initial attempt to achieve this (Baldwin *et al.*, 1987a–c) was partially successful and is one good example of a form of representation which might be further developed to solve the critical issue of nutrient partition. We would point out, however, that a separate representation of control mechanisms which influence the partition of fat precursors between tissues and milk and of protein precursors between tissue and milk may be needed.

As regards prediction of substrate–product relationships at the point of synthesis, conventional stoichiometries may not be sufficient to account for overall nutrient utilization. Modifications to allow for associated pathways of metabolism will be needed and these present a particular challenge. While it may be relatively straightforward to identify the supplementary amounts of glucose which are needed to allow units of acetate to be incorporated into triglyceride, the further linkage between such a process and 'supplementary maintenance' are not so clear. This is an extension of the long-standing issue in the prediction of protein and lipid accretion, in which the energy costs associated with these phenomena exceed those which would be expected from conventional stoichiometry.

A question fundamental to response prediction in the dairy cow is: 'What limits the synthesis of the milk constituents?'. It is self-evident that without an adequate mammary supply of appropriate precursors, the production of milk fat, protein and lactose must be constrained. However, it is much less certain that such constraints normally apply. The mammary gland has been shown to exhibit a rapid response in product output to an increase in precursor supply, with respect to LCFA (Storry *et al.*, 1969), whilst recent data with jugular infusion (Metcalf *et al.*, 1996b) reported large increases in milk protein yield and content when only essential AAs were infused. In the case of LCFA, the relationship between precursor supply and total fat secretion is complicated by the adverse effects of pre-formed fatty acids on the mammary synthesis of short- and medium-chain fatty acids.

There are well-documented examples of lactating cows and goats being stimulated through injections of somatotropin or through increased frequency of milking (Blatchford and Peaker, 1983). In each case, increased milk production was in part a result of repartition of nutrient use away from body tissues towards the mammary gland. These responses indicate that prior to external manipulation, milk production was restricted by a shortage of nutrients due to impaired intake or to increased partition to body tissues, rather than to the capacity of the mammary gland to synthesize milk solids.

In moving towards a prediction of nutrient use, limiting rates of nutrient disposal for particular synthetic (or catabolic) processes will require definition. These can be expected to relate to inherent characteristics of the animal.

7.   Characterization of the Cow

7.1   The need for characterization

Adequate descriptions of the cow, its feed and its environment are essential prerequisites for any satisfactory attempt at response prediction. Chemical descriptions of the feed are reported routinely in the scientific literature. While it is questionable whether current methods for chemical characterization are completely adequate to allow reliable predictions of the nutrient supply which is available to the animal, there has been a long-running and sizeable effort to establish methodologies and a reasonably widespread agreement on the specific nutrients, the supply of which requires to be predicted from feed characteristics. In contrast, quantitative descriptions of animal characteristics and environment are usually less than thorough; an animal's age, weight, parity and stage of lactation might be given but rarely is there a quantitative assessment of the animal's current fatness or stage of growth or maturity (as protein mass) or any indications of its 'potential' for milk production. Nor is there, at present, a clear consensus on the measures which are necessary to define the animal appropriately. The Cornell net carbohydrate and protein system (CNCPS) (Fox *et al.*, 1988, 1992) uses condition score (CS) 1–5 to predict total body fat stores and both total body protein mass and mobilizable body protein (0.25 of total) in the dairy cow. Wright and Russell (1984) provided relationships between CS, weight and body composition of cattle of different breeds, including Holstein–Friesian cattle. More recently, Gibb and Ivings (1993) have developed similar relationships specifically for Holstein–Friesian cows. The general applicability of CS and live weight to predict body composition across genotypes has never been tested (for example, does a CS of 2.5 imply the same body composition for a pure, high-merit Holstein as it would for an average Friesian?). Oldham and Emmans (1989b) outlined a scheme for describing, quantitatively, the lactation potential of cows and this has subsequently been developed into a workable framework (Friggens *et al.*, 1998). Of all the characteristics of cows which demand quantification this is surely paramount, yet its definition has received remarkably little attention compared with that which has been given to descriptions of feed.

There can be little doubt that an animal's inherent characteristics and current physiological state both play a major part in its response to the nutrient supply from the diet. An important aspect of the development of systems for response prediction must therefore be an improved description of cow characteristics. The following sections consider aspects of the description of the cow and discuss approaches that might be taken to incorporate the characterization of the cow into models of response prediction.

7.2   Composition of the body

There is ample evidence that pregnancy and lactation have profound effects on body composition. However, in its extensive review of the composition of the empty bodies of cattle, ARC (1980) came to no firm conclusion on the composition of empty body gain or loss in lactating cattle. Subsequently, a series of slaughter experiments to determine changes in the total body composition of lactating dairy cows was undertaken at IGER, Hurley (Gibb *et al.*, 1992; Ivings *et al.*, 1993). This work led to a conclusion as to the mean energy value (19.3 MJ kg^{-1}) of live-weight change in lactating dairy cows, which was adopted by AFRC (1993). The protein content of live-weight loss was found to be 158 g kg^{-1}, similar to the ARC (1980) estimate of 138 g kg^{-1} based on work with lactating sheep, not dairy cows. There can be a wide range of compositions of body state change. Figure 5.3 in Oldham and Emmans (1989a) shows the estimated protein and fat changes in the bodies of cows used in calorimetric measures of N and C balance (Van der Honing, 1975; and unpublished results). All possible combinations of protein and fat gain and loss are seen. To make such changes predictable is important for the assessment of responses to nutrients.

Pregnancy, over a range of species, results in higher weight gain and N retention than would be expected from the increments in the products of conception, higher blood volume, increases in liver and gut weights and the development of mammary tissue. Lactation further increases blood volume, gut and liver weight (Robinson, 1986). The anabolic effects of lactation on body protein are often associated with cyclical changes in body fat (Johnson, 1973). Some of these changes are temporary but the stimulatory effect of successive pregnancies on growth rate suggests that the endocrine environment can influence mature body weight of female animals. It should be noted, however, that Hovell *et al.* (1977) found that *post partum* losses of N could equal pregnancy anabolism in the pig whilst Johnson (1973) found that the additional N retention associated with pregnancy and lactation in the mouse was not accompanied by an increase in skeletal size. The importance of defining mature body weight in the prediction of potential growth rate has been highlighted by Taylor (1973). The extent to which pregnancy and lactation might influence assessments of mature weight has not been clarified.

None of the effects over successive pregnancies, which have been described above, have been determined in the cow but limited information from within-lactation studies suggests that the cow behaves in a manner analogous to other species. For example, Smith and Baldwin (1974) found that as a result of lactation, there was a redistribution of tissues towards the gut and the udder in the cow and Reid *et al.* (1986) observed increases in liver weight. Many data confirm that body fat is depleted during early lactation (Bath *et al.*, 1965; Reid and Robb, 1971; Chigaru and Topps, 1981; Butler-Hogg *et al.*, 1985) with losses as high as 1.4 kg day^{-1} being recorded (Belyea *et al.*, 1978). Butler-Hogg *et al.* (1985) found that as a result of a loss of dissectable fat of 0.8 kg day^{-1} up to mid-lactation, 0.48 of the cow's fat reserves present at calving were depleted. Gibb *et al.* (1992) reported post-calving empty body contents of fat of 110 kg for multiparous Holstein–Friesian cows. By week 8 of lactation this had declined to 73 kg, a loss of 37 kg (0.7 kg day^{-1}) or 0.34 of that present at calving. Beever *et al.* (1998), working with very high yielding (40–50 kg day^{-1}) cows, has recorded energy deficits of the order of 20–40 MJ day^{-1} up to week 20 of lactation, not related to live weight change, which was sometimes positive. The range of possible body compositions in cows is large, especially for fat (Oldham and Emmans, 1989a). The rates and extents of body fat loss will reflect current fatness (genetic), propensity for milk production and current nutrition, as well as stage of lactation.

Direct evidence from other species and indirect evidence from dairy cows shows that the amount of fat lost during lactation is proportional to the animal's body composition as reflected in its fatness at calving (Cowan *et al.*, 1980; Land and Leaver, 1980; Garnsworthy and Topps, 1982; Fox *et al.*, 1988). Surveys across experiments in which CS was monitored over periods of lactation (Garnsworthy, 1988; Broster and Broster, 1998) suggest that the magnitudes of changes in CS are in some way related to CS at calving. The observations are consistent with a view that there is a 'target' CS which cows seek to achieve if possible. Body CS above the target value allows mobilization, whereas CS below the target is likely to be associated with enhanced food intake and partition of nutrients to favour recovery of CS (Garnsworthy, 1988).

Such target patterns for CS may vary with genotype. Veerkamp *et al.* (1994) have shown that patterns of CS change over the lactation differ between Holstein–Friesians according to pedigree index (for fat + protein kg) and feeding system. Long-term studies (first three lactations) with these genotypes and feed systems have shown that maintenance of a minimum CS may become a constraining factor in the realization of genetic advantage (especially in lower input feeding systems) and this becomes quantitatively more significant over multiple lactations (Chalmers *et al.*, 1997). These data highlight the importance of developing concepts about the meaning of genetic descriptors for body composition in cows at different stages of the lactation cycle as well as for the rates of use of reserves. These comments apply particularly to the fat content of the body. The introduction of ideas about target patterns of fatness is consistent with the new understanding of processes which govern relationships between body fatness and food intake in other species (Emmans and Friggens, 1995). For other species the identification of the protein product of the *ob* gene, leptin (Zhang *et al.*, 1994) has stimulated an explosion of research activity into its function (Trayhurn, 1996).

There is also evidence of a loss of protein reserves during lactation. Botts *et al.* (1979), using a depletion/repletion technique, have calculated that protein reserves in the cow represent around 0.25 of total body protein, although normally losses in early lactation may be only 0.02–0.15 of initial protein weight. Gibb *et al.* (1992) reported post-calving empty body contents of protein of 85 kg for multiparous Holstein–Friesian cows. By week 8 of lactation this had declined to 78 kg, a reduction of only 7 kg (140 g day^{-1}) or 0.08 of the post-calving figure. Limits to the mobilization of protein in cows may rarely be reached but, by analogy with measurements made in lactating rats (Pine *et al.*, 1994), one might expect very significant reductions in lactation output in animals whose readily labile protein reserves are approaching full depletion.

There are modifying effects related to the composition of the diet. Slaughter experiments with sheep indicate that severe restriction of the total diet results in losses of body protein and fat (Sykes and Field, 1974), whilst restriction of energy with adequate protein supply results in the depletion of fat but not of body protein (Cowan *et al.*, 1980).

Few data exist on the body composition of pregnant dairy cows, and these come mainly from indirect determinations. There are few reports of the influence of nutrition on direct measurements of body composition changes in cows. Given the importance of defining the current state of the cow in the prediction of response, there is a clear need for reliable techniques such as ultrasound scanning (Ivings *et al.*, 1993) for *in vivo* assessment of body composition (fat and protein) in cows to be developed and applied, and for critical tests of concepts to establish the role of body composition in determining response relationships.

Despite the lack of information on body composition of the cow a conceptual framework can be stated based on work with other species:

1. Growth of the cow may be described by a Gompertz function and composition described as an allometric function of the degree of maturity, recognizing that mature weight of the female may be influenced by its endocrine environment.

2. Pregnancy and lactation lead to associated cyclical changes in the masses of body protein and fat. In the case of fat there may be a target fatness at a particular stage of lactation which is a function of genotype.

3. The labile reserves of protein are markedly less than those of fat.
4. The ratio of fat to protein in body-weight change depends on the animal's initial fatness and the amount and composition of the diet (i.e. the nutrient supply).

7.3 Influences of the cow on the secretion of milk constituents

Differences in milk yield and in milk fat, protein and lactose contents related to breed of cow, stage of lactation and lactation number are well documented (Waite *et al.*, 1956; Rook, 1961; Crabtree, 1984). Also there are well-established environmental and management effects arising from the frequency and completeness of milking, the length of the milking interval and the occurrence of udder disease (Dodd, 1984). Much of the information available on these effects is summarized in the literature as mean values for groups of animals. However, both milk yield and composition show considerable between-animal variation (Crabtree, 1984), reflecting both inherited traits and differences related to past and present nutritional status and physiological state. While data from recorded herds with pedigree information allows statistical assignment of between-animal variation to genetic and other sources, the biological basis of this individuality is poorly understood and the extent to which the individual characteristics of the animal affect its responsiveness in yield of individual milk constituents to change in diet has not been examined.

As evidenced by the variations in depression of milk fat in response to high-concentrate, low-fibre diets, animal effects related to stage of lactation (Broster *et al.*, 1979) have been shown to exist. But such effects have yet to be systematically investigated and quantified. Indeed, their importance has to a degree been disregarded since it has been assumed, quite wrongly, that responsiveness in milk volume is an adequate index of responsiveness in milk fat and protein yield (Johnson, 1983b).

Variations between individuals, between physiological states within an individual or as a result of differences in nutrition are all, in theory, predictable, subject to an understanding of the nature and controls for the processes which underlie the differences. In order to develop a view of 'nutrient response' an appropriate conceptual framework is needed which gives proper recognition to differences between individuals in animal characteristics which are pertinent to the response of cows to their diet. Such a framework will include a description of the feeds which allows the prediction of the nutrient supply from the diet and a description of the routes of nutrient disposal and their relationships one with another. It will have elements to provide:

1. A description of the individual's potential to yield milk constituents.
2. A description of body composition.
3. A set of rules which relate nutrient supply and nutrient use according to the defined characteristics of the cow and the diet.

7.4   Cow potential and priorities in nutrient use

Dairy cows progress through growth and the cycles of pregnancy and lactation in a controlled manner, regulated by homeorhetic and homeostatic mechanisms. The existence of controls which help to direct change reflect the cow's 'purpose'. For example, once pregnant, a major part of the cow's 'purpose' is to allow the fetus to grow and develop and, towards the end of gestation, to allow the mammary gland to prepare for the secretion of colostrum and then milk.

It is helpful in a description of the dairy cow and her characteristics, to establish a view of the biological path that the cow is trying to follow so as to assess the importance of the nutrient supply on her ability to attain her goals (Emmans and Fisher, 1986). A cow's 'potential for milk production' is frequently referred to, but 'potential' should also be considered in relation to capacity to grow (rate of growth, mature size), capacity to use body tissue reserves in support of milk production, and capacity to eat bulky feeds, at least. It is clearly impossible to predict the cow's response to a change in nutrient supply unless the cow's initial state in relation to her 'potential' can be defined in appropriate quantitative terms. Is there 'potential to respond' or not?

In reality the animal is almost always seeking to achieve several goals simultaneously, but some will be more important than others so there is an order of priority. Hammond (1944) summarized this concept for growing animals using the suggestion of Childs (1920) that nutrient partition between different tissues is determined in order of the metabolic rate of tissues, tissues with high metabolic rate taking precedence over those with lower metabolic rate. But metabolic activity of tissues changes with physiological state (the mammary gland would be an extreme example) and tissue metabolic rate can be regarded both to cause and to respond to changes in the priority rules exercised by the animal.

Truly objective description of a hierarchy of metabolic processes is difficult and may not be absolute – the priority order will change with physiological state, and will reflect the balance between homeorhetic and homeostatic regulation of nutrient use. A general principle, however, would be that maintenance of body integrity would take highest priority, and achievement of fatness the lowest. These principles are implicit within most existing models of nutrient use (e.g. Graham *et al.*, 1976; Gill *et al.*, 1984). For the dairy cow the following could be considered as a first description of the 'purposes' that the cow is seeking to achieve:

1. The cow will give highest priority to those functions which are essential for maintenance of life and metabolic integrity.

2. The cow will aim to maintain a minimum protein mass at a particular stage of maturity (Taylor, 1985) and probably will also aim to maintain a certain fat mass – the latter possibly being relatively small, but the former substantial.

3. If pregnant, the cow will maintain its fetus at the expense of milk production if nutrition is poor.

4. The cow will aim to produce milk, the pattern of production of milk fat, protein and lactose being related to the stage of lactation, age and parity of the cow (Rook, 1961).

5. The cow will aim to achieve an upper limit of body protein mass at a particular stage of maturity. As described above, cows are liable to deplete body protein substantially in lactation under extreme nutritional conditions, but there is an upper limit to protein mass at any stage of maturity. Hence, 'potential' to grow, as defined by mature size, needs to be described with an upper and lower boundary.

6. The cow will aim to achieve a desired fatness according to stage of lactation.

Whether a fixed order of priorities exists or whether priorities can change, subject to homeorhetic drive, is amenable to test by experimentation. For the purpose of this report, it is sufficient to emphasize that to progress the approach of predicting response to nutrient supply it is necessary to define rules for the partition of nutrients between alternative pathways of utilization, and these must sensibly represent the partition which is observed in dairy cows.

8.   Conclusions

On the basis of the information presented in this report the Working Party has concluded that it should be technically feasible to develop a feeding system which allows the yield of milk fat, protein and lactose to be predicted from a knowledge of the nutrient supply furnished by the diet and the metabolism of nutrients by the cow.

The system would be radically different in approach to the current ME and MP systems in that the supply and utilization of intact proteins, amino acids and energy-yielding nutrients (VFA, glucose and LCFA) would be considered within a single framework which would also include elements reflecting the metabolic characteristics of the cow. In concept, the approach would be to predict the amounts of the main end-products of digestion formed in the rumen, small intestine and caecum and colon from the chemical composition and nutritional characteristics of the feeds included in the diet, then to predict the utilization of the end-products of digestion for basal metabolism, fetal growth, synthesis of milk fat, protein and lactose, and synthesis of body protein and body fat. The system would need to be designed in the form of a computer model which would allow interactions between dietary constituents during digestion to be specifically recognized and would facilitate calculations of the partition of nutrient use between competing metabolic processes.

It is clear, however, that the development of a satisfactory new system will require a substantial research effort. The broad approach that should be adopted has already been outlined and major areas where scientific understanding or lack of information limits progress have been identified. This leads to recommendations for future research which are outlined in Chapter 9. Some of these recommendations are of a general nature whilst others are specific; both imply shifts in the emphasis of current research programmes.

9. Recommendations for Research

Our recommendations for the research that will be needed for the new type of feeding system to be developed are summarized under four main headings. However, it is implicit in the general approach that, so far as possible, research falling into the different categories should be integrated to allow maximum benefit to be derived from new findings and new information.

9.1 Feed characterization

There should be a rigorous reassessment of the methods for the characterization of feeds. Research should be directed towards the development of chemical, physical and biological methods that characterize feeds in terms that allow a meaningful description of the dietary inputs to microbial fermentation in the rumen and to enzymic digestion in the intestine (see Section 10.2.1). It follows from this that there should be a positive policy to encourage researchers to provide in published experimental reports a more comprehensive description of feeds than is normal at present. These inputs have been considered by the Working Party on Feed Characterization (AFRC, 1987a,b, 1988) and updated by Givens (1994) as shown in Fig. 9.1. However, it is important to recognize that further modifications to systems of feed characterization will need to be made to accommodate new approaches to the prediction of response.

9.2 Nutrient supply

There should be continued emphasis on research directed at establishing the dietary and animal factors influencing the formation of the products of digestion. Such studies should be designed mainly to elucidate the mechanisms regulating the processes and to provide information on responses to

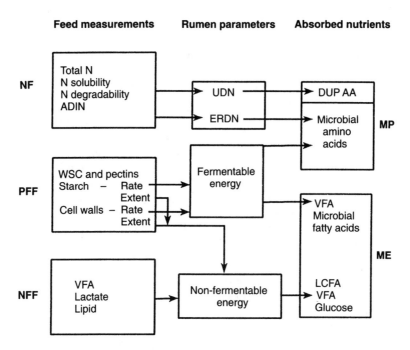

Fig. 9.1. Possible future model for feed characterization (after Givens, 1994). AA, amino acids; ADIN, acid detergent insoluble N; CW, cell walls; DUP, digestible undegradable protein; ERDN, effective rumen degradable N; FA, fatty acids; LCFA, long-chain fatty acids; ME, metabolizable energy; MP, metabolizable protein; N, nitrogen; NF, nitrogen fraction; NFF, non-fermentable fraction; PECT, pectins; PFF, potentially fermentable fraction; UDN, undegradable N; VFA, volatile fatty acids; WSC, water-soluble carbohydrates.

dietary change. There are particular limits to knowledge of the regulation of:

- microbial fermentation and growth in the rumen and how this relates to microbial yield to the abomasum;
- VFA production in the rumen;
- lipid digestion.

These topics are of central importance with respect to dietary effects on milk production in the dairy cow.

Studies on digestion and nutrient supply should not be undertaken in isolation however, for it is important that they are linked with experiments that provide an assessment of nutrient use. Similarly, the animals, diets and levels of feeding used for experimentation should be relevant to the prediction of responses of lactating dairy cows under commercial conditions.

9.3   Nutrient use

A major new initiative is needed to improve understanding of quantitative aspects of the metabolism of absorbed nutrients in the dairy cow, instead of only considering energy as ME and either MP or absorbed AAs, as in the CNCPS and INRA models. Work to identify the efficiencies of utilization of each nutrient for each synthetic purpose should be encouraged. These studies must recognize the importance of variations in the nutrient supply and nutrient interactions both within the mammary gland and extra-mammary tissues and should encompass nutritional–endocrine effects on the partition of nutrient use between competing metabolic processes. At present there is a substantial gulf between knowledge of the biochemical pathways of metabolism and scientific appreciation of the quantitative significance of those pathways in the whole animal. In the context of this report, much greater emphasis should be given to whole-animal studies with relevant animals in appropriate productive states, although carefully designed *in vitro* experiments can still make a contribution.

For the required progress to be made there will need to be significant developments in the methodology of *in vivo* experimentation. In particular, techniques are needed to determine the quantities of feedstuffs absorbed by the gastrointestinal tract, the magnitude of endogenous secretions produced, and the efficiencies with which the individual products of digestion are used for maintenance. The accuracy of A–V difference techniques, and of continuous blood flow measurements in particular, must be improved and innovative approaches to isotope tracer techniques are required to overcome some of their present limitations.

9.4   Characterization of the cow

An essential feature of the proposed system for nutrient response prediction is the inclusion of the functions to describe the cow's current 'state' and 'potential' for synthesis of milk constituents, body protein and body fat. This concept is a major departure from the approach used in the majority of existing ruminant 'nutrient' requirement systems based purely on factorial principles for the estimation of these requirements, and needs to be fully and critically evaluated.

Information is required on the rules which govern body composition of dairy cows at different stages of growth, pregnancy and lactation, and on the changes in body composition which occur in response to alterations in diet and nutrient supply.

Additionally, however, the problems of biological definition and scientific measurement of animal 'potential' for the secretion of milk constituents, for body protein synthesis and for fat deposition have remained largely unaddressed. Given that the udder is the major site of synthesis and storage of milk solids, it is surprising that so few measurements have been made of its anatomical structure in relation to the potential to yield milk, although Dewhurst and Knight (1993) have investigated phenotypic variations in sites of milk storage in the udder. For the practical application of any model which depends on a description of genotype it will be necessary to identify measures which can be taken or indexed readily in practice. The extent to which conventional measures of genetic worth (either alone or in combination with other accessible measures) can be used for this purpose needs to be evaluated.

The realization of genetic potential for milk production may be limited by the capacities of cows to ingest and process feeds under some circumstance, and this might apply especially to the capacities of cows for 'bulk' characteristics of feeds. Research to evolve rules to describe the capacities of cows for bulk intake will also be valuable. In many respects it is these problems which represent the most scientifically challenging areas of research in the development of feeding systems designed for response prediction.

10. Approaches to the Prediction of Response in Milk Constituent Output

10.1 Modification of the ME system

A number of attempts have been made (Blaxter and Ruben, 1953, 1954; ARC, 1965, 1980; Blaxter, 1967; Hulme et al., 1986) to use existing energy requirement systems as a basis for the development of dynamic models to predict responses in milk energy output to changes in dietary energy supply. The emphasis in all the studies has been to describe rules which define the partition of ME between body energy deposition and milk energy secretion.

Early attempts to model response adopted the approach that responsiveness was directly related to the state of the cow, defined in terms of the current milk yield at a particular plane of nutrition. Thus, Blaxter and Ruben (1953, 1954) and Burt (1957) suggested that response, in terms of additional yield of milk energy to an increment of feed energy, was a function of the current yield of the cow fed to the SE standard. Partition of an increment of feed energy between milk and use for body tissue synthesis was then, implicitly, a consequence of this rule. Amended versions of this approach, converted from SE units to MJ of ME, were included in the dairy cow energy requirements section of ARC (1965) and ARC (1980), but were not adopted for use or tested further. The model of Hulme et al. (1986) has also used a similar approach. Hulme and his colleagues re-analysed the response curves for cows fed at six planes of nutrition (Jensen et al., 1942), in which recorded levels of milk yield were only 10–20 kg day^{-1}. Hulme et al. (1986) concluded that the efficiency of milk energy secretion declined as energy intake was increased and that the rate of decline in incremental response was greater in cows of low current yield than those of high current yield.

The concept of defining cow state in terms of current yield has been extended by Wood (1979) to consider both the capacity to yield milk and the potential to deplete and replete body reserves. If it is assumed that the transformation of dietary energy to milk energy and to body energy and the transfer of body energy reserves to milk energy all occur with specific and fixed efficiencies then the framework can be derived in which dietary energy is divided between milk production and live-weight gain in proportion to their demands. Within this context, potential demand is described according to an assumed lactation curve whilst potential for tissue mobilization, or deposition, is related to a reserve of tissue energy (R). Thus, the rate at which this reserve could be used to support milk production is described by:

$$\frac{\mathrm{d}(R_n)}{\mathrm{d}t} = -k^*R^*\mathrm{e}^{-kn}$$

where k is a rate constant and n is the week of lactation.

There is some validity in this approach since, as was indicated earlier, the extent of fat loss in early lactation seems to be related to the size of the fat reserves present at the time of calving. However, predictions using this model suggest that tissue energy loss in early lactation is obligatory unless the potential milk energy output is achieved within the limits of the dietary energy intake. Furthermore, because of the fixed partition between milk energy secretion and tissue energy gain which is assumed at all stages of lactation, live-weight loss in later lactation would not be predicted. Also, as the parameters of the assumed lactation curve are derived from observed data on milk yield the curve may reflect the cow's current rather than her potential yield of milk.

The approach of Wood (1979) has been extended by Bruce *et al.* (1984) with a modification of the concept of potential which allows growth of the cow's body to be separate from fat deposition. During a period of energy shortage, milk production and maternal body change are reduced by equal amounts relative to their potential. Energy consumed in excess of both milk and growth potential is assumed to be deposited solely in the form of fat. This approach has been included in the Cornell CNCPS model (Fox *et al.*, 1992).

The authors of the models outlined above have made use of a variety of mathematical and biological concepts but all have adopted a framework incorporating some or all of the following assumptions:

1. Maintenance and then the requirements of the developing fetus make a first charge on the energy supply, the excess being available for milk production and body tissue synthesis.
2. Provided that there is no change in total body energy, milk energy output is directly proportional to the energy available, and the parameters of the relationships are independent of the cow and the stage of lactation.
3. For each lactation cycle, each cow has maximum achievable coordinates on the energy intake/milk production curve. These define the cow's capacity to produce milk during that lactation cycle, reflecting the cow's 'potential' for milk production within the constraints imposed by environmental and management factors.

4. Additional to cyclical changes in total body energy over lactation, the cow has a preferred rate of growth which is a function of its mature body weight.

Departing from the point at which the cow neither loses nor gains body energy, increments of the energy intake are partitioned between milk production and synthesis of body tissues. Such responses have two main features. Firstly, there is a lag between input and output which depends on whether tissue energy is being gained or lost. Secondly, the parameters of the response curve depend on the capacity of the cow to yield milk, to grow and to deplete and replete body reserves.

Although the authors of the various published models provide to a greater or lesser extent evidence of their accuracy of prediction, the models have not been subjected to rigorous tests against randomly selected sets of data. Moreover, it has to be recognized that the models seek only to predict milk energy output. They do not attempt to partition milk energy between milk fat, protein and lactose since it is assumed that milk composition is solely an attribute of the cow, as in the CAMDAIRY and CNCPS models (Hulme *et al.*, 1986; Fox *et al.*, 1992). Nevertheless, as a first step in the development of systems to predict responses in milk constituent output it would appear sensible to examine whether the published models of energy utilization (e.g. Van Es, 1978; ARC, 1980; INRA, 1988; NRC, 1988; AFRC, 1990) are amenable to modification. Clearly such modification would need to be applied subsequent to the prediction of milk energy and would need to take account of the need to define the diet not only in terms of ME and energy concentration M/D (q_m), as in the original ME scheme of ARC (1965), but also in terms of its detailed chemical composition. It should not be forgotten, however, that variations in diet M/D describe in broad terms the shift from fibrous, low starch, low (7–10 MJ kg^{-1} dry matter (DM)) M/D forages to low fibre concentrate feeds high in starch and protein, with high (11–13 MJ kg^{-1} DM) M/D values. The original explanation of the variation in the efficiency of fattening was the shift in the acetate proportions in the VFA produced and absorbed in the rumen (Blaxter and Wainman, 1964). Efficiency of utilization of ME for milk synthesis (k_l) was similarly found to be related to M/D (q_m) (Moe *et al.*, 1972; Van Es *et al.*, 1978).

In their simplest form the modifications might consist of regression equations incorporating empirical descriptions of established relationships between diet composition and milk composition (e.g. relationships based on fibre content, fibre:starch ratios, etc.). Furthermore, the accrual of information on solids output over the lactation permits the continuous correction of the response prediction given a framework which allows a delay between food consumption and milk secretion (Evans *et al.*, 1985).

This approach of modifying the ME system, although clearly amenable to test, has important limitations. Primarily it is implicit that the characteristics of the feed have an influence solely on the partition of milk energy between fat, protein and lactose but, as discussed earlier, the partition of nutrients between milk secretion and body tissue deposition is also sensitive to the composition of the mixture of nutrients derived from the diet.

10.2   Towards a new approach

As discussed in Chapter 4 of this report, a satisfactory representation or framework for prediction of milk production responses in dairy cows will need to be dynamic, and should be as simple as possible with due regard to the need to meet the stated objectives. It is logical to conceive of the framework as having one major element representing nutrient supply (digestion of the diet) and a second representing nutrient utilization, with transactions occurring in the context of a description of the cow in terms of its current state, as outlined in Fig. 4.1.

10.2.1   Prediction of the end-products of digestion

There have been several attempts to predict the end-products of ruminal digestion using empirical equations (Hogan and Weston, 1970; Beever *et al.*, 1976; Ulyatt and Egan, 1979; Russell *et al.*, 1992; Sniffen *et al.*, 1992; Pitt *et al.*, 1996) but the limitations of such approaches have been identified especially with respect to their inability to provide reliable predictions over a wide range of diet types. This point was elaborated by McAllan *et al.* (1987) in relation to the effect of diet type on the efficiency of RDN utilization and the yield of microbial protein. Accurate prediction of both the amounts and proportions of the VFA absorbed from the rumen have also proved difficult (Baldwin *et al.*, 1987b; Dijkstra *et al.*, 1993; Pitt *et al.*, 1996), which, given their significance as the major portion of the energy supply to the cow, and their effects upon milk synthesis and hormonal status, must clearly be clarified.

On the other hand considerable progress has been made in the development of simulation models of rumen fermentation (Baldwin *et al.*, 1970, 1977, 1987b; Black *et al.*, 1980/81; France *et al.*, 1982; Dijkstra *et al.*, 1992, 1993, 1997; Dijkstra, 1994; Dijkstra and France, 1996) although it is reasonable to conclude that these represent research models and at this stage it would be inappropriate to consider them as robust predictive models. Such approaches permit integration of the many varying, and often competing, processes which occur within the rumen and have the advantage of allowing time to be explicitly represented. The procedures for the construction of such models has been extensively reviewed elsewhere (France and Thornley, 1984) and will not be discussed in any detail here. In principle, the models deal with transformations between inputs to chemical or biochemical reaction sequences and outputs of end-products. This is achieved by describ-

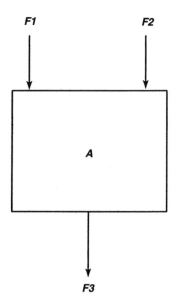

Fig. 10.1. A representation of a simple model.

ing the key intermediate stages of reaction sequences in terms of the amounts of intermediates present (a mass, represented schematically as a box) and the rates of flow of substances into and from the intermediates (masses per unit time, represented schematically by arrows). Thus, as an example, Fig. 10.1 illustrates a reaction in which the flow of two substances (*F1* and *F2*) combine via an intermediate stage *A* to give a flow of product *F3*.

The mass of *A* (the amount represented within the box) will change with time depending on the relative magnitudes of *F1*, *F2* and *F3*, and over a given time interval will be described by the differential equation:

$$\frac{dA}{dt} = (F1 + F2) - F3.$$

Within a more complex model the number of inputs and outputs for a particular 'box' can vary, as can the rate functions which influence the flows of substances between 'boxes'. Moreover, some rate functions can be variables which depend on the status of the other components within the model and interaction effects can be represented in mathematical form by describing relative affinities for nutrients between different pathways or processes. Frequently this approach has been based on equations of the Michaelis–Menten form to describe the relationship between actual rate (*R*; mass per unit time) of a process (conversion of substrate, *S*, to product, *P*) and the maximum possible rate for that process (V_{max}). Available substrate [S] and an affinity constant (k) which is a measure of the affinity of substrates for the conversion process, need to be defined. Thus:

$$R = \frac{V_{max} * [S]}{(k + [S])}$$

The relative contribution of different potential precursors of P (say S_i) will depend not only on available quantities of S_i but also on defined values for V_{max} and k for each of the S_i to P conversions. Similarly, use of S for the synthesis of different products (P_i) will also depend on V_{max} and k values for the various S to P_i conversions. Partition of the substrate S between competing pathways therefore depends on relative values of V_{max} and k, for the different processes or reactions. With this approach there will always be some transfer of S to each P_i (although a reverse reaction might result in no net conversion of S to P). Other factors influencing the conversion of S to P can also be allowed for by modifying the form of the Michaelis–Menten equation by including additional terms. This approach has been used, for example by Gill *et al.* (1984), and forms the basis of the model given in the appendix.

To predict the response of the cow to changes in nutrient supply, an estimate of the supply of the following nutrients arising from ruminal and intestinal digestion may be a minimum requirement:

- acetate,
- propionate,
- butyrate,
- glucose,
- AAs,
- LCFA.

In order to predict the supply of these nutrients, certain chemical and physical characteristics of the feeds need to be defined. These are outlined below, together with some assumptions regarding their modes of digestion. Figure 10.2 provides a schematic 'model' representation of digestion in the rumen, and the appendix provides a structured set of propositions by which it may be possible to describe digestion in the rumen, small intestine and caecum and colon. At this stage, these propositions are not regarded as definitive statements about the way in which the relationship between diet and nutrient supply should be expressed. Rather this 'nutrient supply model' represents a considered view of how the problem of predicting nutrient supply should be approached. As such, the model represents a framework which could form the basis for future development.

Within the model, the inputs have been identified with the major chemical components which are present in the diet:

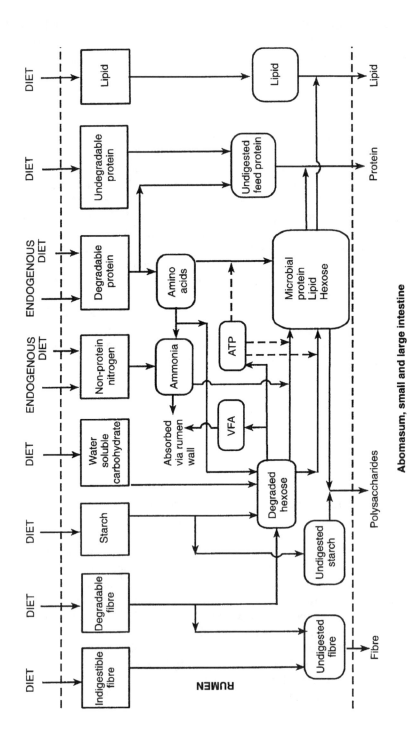

Abomasum, small and large intestine

Fig. 10.2. A schematic representation of a model of digestion of dietary constituents in the rumen.

- starch,
- water-soluble carbohydrates,
- pectins,
- fermentation acids,
- fibrous carbohydrates,
- protein nitrogen,
- NPN,
- lipid.

These selected components represent a compromise between a more comprehensive description of feed, which could be justified on theoretical grounds, and the more limited description which is often available in practice. Especially for the fibrous carbohydrates, starch and protein, it seems likely that current chemical descriptions of feeds will prove inadequate for predictive purposes, since estimates of likely rates of degradation of each nutrient fraction when in contact with rumen bacteria are required by all current models. In this context, the gas pressure transducer *in vitro* rumen liquor technique (Theodorou *et al.*, 1994) shows promise. This should provide a challenge to those involved in feed characterization to refine and extend their methodologies. This follows in part from a view that the laboratory characterization of feeds will need to take account of the present purpose of constructing an illustrative model.

The main assumptions that have been made about the characteristics of the feed constituents are as follows:

1. Water-soluble carbohydrates (and pectins) are completely rumen degradable for all diets.
2. Pre-formed fermentation acids can be measured separately and any VFA present will be absorbed as such. The proportions of lactic acid (often present in amounts up to 10% of forage DM) utilized by rumen microbes or absorbed unchanged, need to be described.
3. Both the amount of rumen-degradable starch can be measured, as by Nocek and Tamminga (1991), and also its rate of degradation.
4. The potential availability of the fibrous carbohydrate fraction for microbial fermentation can be defined, together with its rate of degradation in the rumen. (The undigested fraction is assumed to leave the rumen with a rate constant which must be declared.)
5. NPN is assumed to be completely degraded in the rumen with a rate constant of availability which must be declared.
6. Potentially degradable and undegradable fractions of dietary protein must be estimated together with their rates of digestion and removal by outflow from the rumen.
7. Lipids (triglycerides) are assumed to be totally hydrolysed in the rumen but otherwise this fraction is assumed to be biologically inert with regard to the rumen fermentation.

The manner in which the chemical constituents in the diet are related to nutrient supply is shown in Fig. 10.2, together with the major transformations that are considered to occur. In addition to the feed constituents already defined the figure includes terms for ammonia, VFA and microbial biomass that are formed as end-products of fermentation, ATP which is used as a ubiquitous currency for energy transformations and AAs which are represented as reaction intermediates.

The approach as presented has limitations, particularly in relation to the failure to incorporate a component reflecting the significance of the physical form of feeds, which is known substantially to influence outflow of particulate matter from the rumen. For example, in the CNCPS, the parameter effective NDF (eNDF), as defined by Mertens (1985), is used to predict rumen pH (Pitt *et al.*, 1996) and its consequent effects upon rumen microbial growth and rate of carbohydrate degradation. Similarly, the reported suggestions that dietary lipids may affect ruminal digestion of carbohydrate components have been ignored. These and other issues are recognized as matters of importance and will need to be examined further, but at this stage we are aim to illustrate a possible general approach which, if adopted, would require data to provide parameter values for the model.

10.2.2　Prediction of nutrient use

Whilst there have been numerous attempts to develop models of rumen function, there has been considerably less progress with respect to mathematical modelling of nutrient utilization in the tissues. Oldham and Emmans (1988) outlined a framework of nutrient utilization for the dairy cow, based on a hierarchy of nutrient use. This approach involves each important nutrient being considered in turn in relation to defined pathways of use. A nutrient is used for each defined pathway in order of priority. The approach is dynamic in that it represents rates of nutrient disposal. However, it requires the description of a unique order of priority for the disposal of a particular potentially 'limiting' nutrient, allowing for the fact that some pathways of disposal are interdependent. For example, nutrient catabolism for energy generation will depend on the nature and magnitude of synthetic reactions. Shortage of a key nutrient will result in a sharp 'cut-off' in predicted performance, somewhat analogous to the approach used by Fisher *et al.* (1973) for predicting the performance of laying hens to variation in AA intake. Oldham (1987) has suggested that the utilization of AAs for milk synthesis is likely to follow a similar pattern in the dairy cow. Models of AA utilization in dairy cows have been published (O'Connor *et al.*, 1992; Rulquin *et al.*, 1993). The CNCPS AA sub-model of O'Connor *et al.* (1992) adopts fixed efficiencies of utilization for each AA for tissue and milk synthesis, without any consideration of metabolic transformations in the tissues, or possible deamination in the liver to produce energy sources. The model of Rulquin *et al.* (1993) is based on a nutrient response approach for identified limiting AA, in this case, lysine and methionine. They reported a curvilinear response function for milk protein yield to graded increases in the AA concentration in the supply of absorbed AA (estimated as PDI), in contrast with the Fisher (1973) model.

For other nutrients – in particular, perhaps, the partition of the fat precursors between fat accretion in body tissue or secretion in milk – it appears that both processes occur simultaneously and that the priority order for nutrient use can change, in which case a strict hierarchy would not apply.

Gill *et al.* (1984) proposed a research model to describe energy utilization in growing lambs, based on specific representations of individual nutrients and nutrient interactions, and the potential of this approach to improve our understanding of nutrient utilization for growth was demonstrated in the publication of Black *et al.* (1987a,b). Subsequently, Baldwin *et al.* (1987a) proposed a similar model of nutrient utilization in the dairy cow, and when coupled with a digestion model (Baldwin *et al.*, 1987b), a combined digestion/ metabolism model for dairy cows was developed (Baldwin *et al.*, 1987c) capable of examining both short- and long-term effects of nutrition on lactation performance. However, whilst Baldwin *et al.* (1987c) were able to demonstrate the general behaviour of the model, when it was applied to specific situations such as the low milk fat syndrome, deficiencies in the model were identified. Thus it is concluded that whilst many of the concepts included in the model are useful, it does not, in its present form, represent a robust model capable of predicting nutrient utilization and, in particular, nutrient partitioning in all situations. An updated mathematical model of the dairy cow has been published recently by Baldwin (1995) and a user-friendly version (AAMOLLY) for advisers was presented at ADSA '97 (Maas *et al.*, 1997).

However, it remains the opinion of the Working Party that considerable effort will be required to produce a satisfactory model of metabolism in the dairy cow. The present state of knowledge on the nutritional and hormonal interactions is considerably inferior to the knowledge which exists with respect to the processes of ruminal and intestinal digestion, and thus to develop a satisfactory model of metabolism requires research not only in model construction but also in the physiology and biochemistry of animals. It is likely, however, that the best progress will be achieved if both types of research are undertaken in parallel. Models similar to that produced by Baldwin *et al.* (1987c) have already identified areas where more information is required and hence provided an impetus for research effort. Subsequent generations of models which utilize such information should be developed, and through this approach the overall objective is achievable.

The ultimate aim, once all relevant information is available, will be to summarize nutrient utilization by means of a restricted number of variables, using a description of substrate supply as a common currency between a supply and a utilization model. To establish this link it may be necessary to account for the extensive conversion of ruminally absorbed butyrate to 3-hydroxybutyrate which is known to occur in the rumen mucosa and liver and the corresponding conversion of significant but variable amounts of propionate to glucose in the liver. Furthermore, there may be a need to represent obligatory intestinal metabolism of specific nutrients, including AAs and glucose absorbed from the gastrointestinal tract.

At this stage it may be sufficient to consider the main substrates available for extrahepatic metabolism, acetate, glucose, LCFA and AAs. Since for many purposes, 3-hydroxybutyrate is converted to acetate before metabolism, these compounds can often be regarded as equivalent.

Given these constraints, it is possible to construct a mathematical model of nutrient utilization within the tissues using essentially the same approach as that discussed earlier for the prediction of nutrient digestion and supply. The actual approach adopted, however, will remain a matter of debate amongst individuals. In the majority view of the Working Party there are scientific and technical justifications for describing the major processes of nutrient utilization in the cow's body and the mammary gland separately, as represented in Fig. 10.3.

These schematic representations, whilst recognizing the importance of both body and mammary metabolism, do not provide any representation of fetal demands. Finally, it is likely that metabolic regulation in the mammary gland will need to be represented more explicitly. A detailed model of the biochemistry of milk synthesis has been constructed by Baldwin and his colleagues (Baldwin and Bauman, 1984) and some of the concepts they have used could be applied to refining the model outlined in Fig. 6.1 and presented in Chapter 6. It has already been mentioned that parallel development of research type models and experimental research is likely to be the most effective means of achieving the desired objective of a robust predictive model. At this stage it is appropriate to add that it may be possible to use information obtained in this way to refine some of the concepts which were suggested in Chapters 5–7. This could occur in advance of the establishment of a suitable predictive model as defined above, and this adds considerable support to the need to consider mathematical modelling and future experimental research in parallel in future research designed to improve the predictability of response of lactating dairy cows to dietary change.

10.2.3 *Description of cow state and potential to yield milk*

As mentioned before, in any model of lactation performance, it will be necessary to provide some representation of cow state and lactation potential.

From a knowledge of both current (P) and mature body protein (P_m) mass, the stage of maturity of the cow (u) can be represented by:

$$u = P/P_m.$$

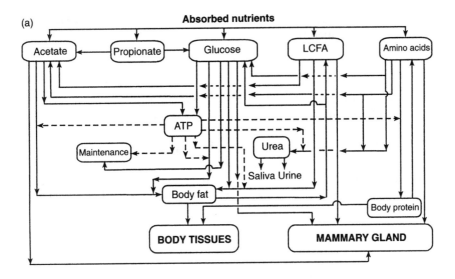

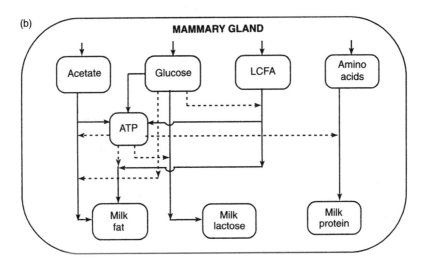

Fig. 10.3. A schematic illustration of a model representing the utilization of nutrients: (a) in the tissue of a non-pregnant dairy cow; (b) in the mammary gland of a non-pregnant dairy cow.

If there is both an upper and lower bound to body protein mass, at any particular stage of maturity, then:

$$P = u*P_m + S_{max},$$

where S_{max} is the maximum 'labile' protein weight, which may be a function of degree of maturity thus:

$$S_{max} = S*u*P_m,$$

where S is the amount of labile protein as a proportion of the non-labile protein.

The cow can be seen as 'seeking to achieve' a particular body lipid mass (L) at a definite stage of maturity (u). This is given by:

$$L = (L_m/P_m)u^b,$$

where L_m is lipid mass at maturity, and b is the exponent relating the degree of maturity to that of protein.

The potential rate at which a cow approaches maturity might be defined by the following relationship:

$$\frac{du}{dt} = B*u*\ln(1/u)$$

where B is the genetic rate parameter. In this approach, it is important that only current and potential state need to be considered in relation to the cow's response to a change in nutrient supply. This is entirely consistent with the role of homeostatic and homeorhetic mechanisms as the main regulators of nutrient partitioning. There may, however, be a need to recognize the impact of growth pattern on mammary development and subsequent performance. Both Sejrsen *et al.* (1983) and Johnson *et al.* (1986) have attended to this, implying that there may be a decrease in the rate of allometric mammary growth in heifers which have exhibited high rates of growth in the pre-pubertal phase. Early attainment of precocious puberty within the major phase of mammary allometric growth may truncate this phase of mammary growth, and override the prevailing hormonal stimuli that promote mammary tissue proliferation. In such respects, it may become necessary to describe mammary growth separately from body protein and lipid mass.

10.2.4   Potential to yield mass

Many empirical models have been proposed to describe the lactation curve, with that proposed by Wood (1976) frequently used as a reference point, namely:

$$Y_t = a*t^b*e^{-ct}.$$

This has the undesirable characteristic of being zero when $t = 0$, whereas it is well recognized that mammary function is well developed at the time of parturition. An alternative, improved representation provided by Emmans and Fisher (1986) is:

$$Y_t = a*u*e^{-ct}$$

where $$u = \exp(-\exp(G_o - B_1*t))$$

The variable (u) is a Gompertz function to describe the development of the mammary secretory function of the cow, G_o describes the state of the mammary gland at calving and B_1 is a genetic constant for lactation. This allows Y_t to be positive when $t = 0$.

Equations of this type have been applied to the secretion of milk constituents, and include both a scaling factor (a) and a function (e^{-ct}) which describes the rate of decline of secretory capacity.

Neal and Thornley (1983) attempted to describe the lactation cycle of milk production using a mechanistic model of mammary gland function. They postulated a 'lactational' hormone which acted to promote milk secretion potential through its effect on cell division. Thus, the rate of mammary gland 'maturing' was a reflection of increased secretory cell number, whilst decay in secretory activity with advancing lactation was effected via a decline in the level of postulated hormone. The capacity of the gland to hold milk was identified as a constraint on milk production, but provided milk was withdrawn from the gland sufficiently frequently, then its milk-holding capacity was not reached.

Thus in theory, with a non-limiting milking regime and a non-limiting and, ideally, balanced supply of nutrients to the cow, the lactation potential of the cow should be achieved. In reality this may or may not coincide with the maximum capacity of the mammary gland to synthesize milk fat, protein and lactose.

To provide adequate descriptions of cows, in support of development of satisfactory nutrient response models, is, however, extremely difficult. The cow's ability to synthesize individual milk constituents, her capacity to synthesize or mobilize tissue fat and protein, her propensity to change feed intake at times of energy deficit or surplus and other such attributes will need to be evaluated. This can only be achieved through a coordinated research effort. Above all, lactation potential will need to be defined, but its measurement in practice may be impossible. As a consequence, indices of responsiveness as proposed by Westgarth *et al.* (1981), or yield in early lactation (e.g. week 2) may be all that are available.

References

AFRC (1987a) Characterisation of Feedstuffs: Energy. Report No. 1, Technical Committee on Responses to Nutrients. *Nutrition Abstracts and Reviews, Series B* 57, 507–523.

AFRC (1987b) Characterisation of Feedstuffs: Protein. Report No. 2, Technical Committee on Responses to Nutrients. *Nutrition Abstracts and Reviews, Series B* 57, 714–736.

AFRC (1988) Characterisation of Feedstuffs: Other Nutrients. Report No. 3, Technical Committee on Responses to Nutrients. *Nutrition Abstracts and Reviews, Series B* 58, 548–571.

AFRC (1990) The Nutritive Requirements of Ruminant Animals: Energy. Report No. 5, Technical Committee on Responses to Nutrients. *Nutrition Abstracts and Reviews, Series B* 60, 729–804.

AFRC (1991) The Voluntary Intake of Cattle, Report No. 8, Technical Committee on Responses to Nutrients. *Nutrition Abstracts and Reviews, Series B* 61, 815–823.

AFRC (1992) Nutritive Requirements of Ruminant Animals: Protein. Report No. 9, Technical Committee on Responses to Nutrients. *Nutrition Abstracts and Reviews, Series B* 62, 787–835.

AFRC (1993) *Energy and Protein Requirements of Ruminants*. CAB International, Wallingford.

Alderman, G. (1987) Comparison of rations calculated in the different systems. In: Jarrige, R. and Alderman, G (eds) *Feed Evaluation and Protein Requirement Systems for Ruminants*. CEC, Luxembourg, pp. 283–298.

Alderman, G., Griffiths, J.R., Morgan, D.E., Edwards, R.A., Raven, A.M., Holmes, W. and Lessells, W.J. (1973) An approach to the practical application of a metabolisable energy system for ruminants in the UK. In: Swan, H. and Lewis, D. (eds) *Nutrition Conference for Feed Manufacturers*, Vol. 7. Butterworths, London, pp. 37–78.

Alderman, G., Broster, W.H., Strickland, M.J. and Johnson, C.L. (1982) The estimation of the energy value of live-weight change in the lactating dairy cow. *Livestock Production Science* 9, 665–673.

Andrews, S.M., Waldo, D.R. and Erdman, R.A. (1994) Direct analysis of body composition of dairy cows at three physiological stages. *Journal of Dairy Science* 77 (10), 3022–3033.

Annison, E.F. (1983) Metabolite utilization by the ruminant mammary gland. In:

Mepham, T.B. (ed.) *Biochemistry of Lactation.* Elsevier, Amsterdam, pp. 399–436.

Annison, E.F., Bickerstaffe, R. and Linzell, J.L. (1974) Glucose and fatty acid metabolism in cows producing milk of low fat content. *Journal of Agricultural Science (Cambridge)* 82, 87–95.

Antoniewicz, A.M., Kosmala, I., Hvelplund, T. and Van Vuren, A. (1998) Use of pepsin–pancreatin system to estimate intestinal digestibility of protein in ruminant feeds and undegraded residues. In: Deaville, E.R., Owen, E., Rymer, C., Adesogan, A.T., Huntington, J.A. and Lawrence, T.L.J. (eds) *In vitro Techniques for Measuring Nutrient Supply to Ruminants.* British Society of Animal Science Occasional Publication No. 22, in press.

ARC (1965) *The Nutrient Requirements of Farm Livestock*, No. 2, *Ruminants.* HMSO, London.

ARC (1980) *The Nutrient Requirements of Ruminant Livestock.* Commonwealth Agricultural Bureaux, Farnham Royal, Slough.

ARC (1981) *The Nutrient Requirement of Pigs.* Commonwealth Agricultural Bureaux, Farnham Royal, Slough.

ARC (1984) *The Nutrient Requirements of Ruminant Livestock*, Suppl. No. 1. Commonwealth Agricultural Bureaux, Farnham Royal, Slough.

Backwell, F.R.C., Metcalf, J.A., Bequette, B.J., Lobley, G.E., Beever, D.E. and MacRae, J.C. (1994) A possible role for peptides as precursors for milk protein synthesis. *Journal of Dairy Science* 77 (Suppl. 1), 1176.

Baldwin, R.L. (1995) The program MOLLY.CSL. In: Baldwin, R.L. (ed.) *Modelling Ruminant Digestion and Metabolism.* Chapman and Hall, London, pp. 476–515.

Baldwin, R.L. and Bauman, D.E. (1984) Partition of nutrients in lactation. In: Baldwin, R.L. and Bywater, A.C. (eds) *Modelling Ruminant Digestion and Metabolism.* University of California Press, Davis, California, pp. 80–88.

Baldwin, R.L., Lucas, H.L. and Cabrera, R. (1970) Energetic relationships in the formation and utilisation of fermentation end products. In: Phillipson, A.T. (ed.) *Physiology of Digestion and Metabolism in the Ruminant.* Oriel Press, Newcastle upon Tyne, pp. 319–334.

Baldwin, R.L., Koong, L.J. and Ulyatt, M.J. (1977) A dynamic model of ruminant digestion for evaluation of factors affecting nutritive value. *Agricultural Systems* 2, 255–288.

Baldwin, R.L., France, J. and Gill, M. (1987a) Metabolism of the lactating cow. I. Animal elements of a mechanistic model. *Journal of Dairy Research* 54, 77–105.

Baldwin, R.L., Thornley, J.H.M. and Beever, D.E. (1987b) Metabolism of the lactating cow. II. Digestive elements of a mechanistic model. *Journal of Dairy Research* 54, 107–131.

Baldwin, R.L., France, J., Beever, D.E., Gill, M. and Thornley, J.H.M. (1987c) Metabolism of the lactating cow. III. Properties of mechanistic models suitable for evaluation of energetic relationships and factors involved in the partition of nutrients. *Journal of Dairy Research* 54, 133–145.

Bath, D.L., Ronning, M., Meyer, J.H. and Lofgreen, G.P. (1965) Calorific equivalent of live-weight loss of dairy cattle. *Journal of Dairy Science* 48, 374–380.

Bauman, D.E. and Currie W.B. (1980) Partitioning of nutrients during pregnancy and lactation. A review of mechanisms involving homeostasis and homeorhesis. *Journal of Dairy Science* 63, 1514–1529.

Bauman, D.E. and Elliot, J.M. (1983) Control of nutrient partitioning in lactating ruminants. In: Mepham, T.B. (ed.) *Biochemistry of Lactation*. Elsevier Science Publishers, Amsterdam, pp. 437–468.

Bauman, D.E. and McCutcheon, S.N. (1984) The effects of growth hormone and prolactin on metabolism. In: Milligan, L.P., Grovum, W.L. and Dobson, A. (eds) *Control of Digestion and Metabolism in Ruminants, Proceedings of the 6th International Symposium on Ruminant Physiology*. Prentice-Hall, Englewood Cliffs, New Jersey, pp. 436–455.

Beever, D.E., Thomson, D.J. and Cammell, S.B. (1976) The digestion of frozen and dried grass by sheep. *Journal of Agricultural Science (Cambridge)* 86, 443–452.

Beever, D.E., Gill, M., Dawson, J.M. and Buttery, P.J. (1990) The effect of fishmeal on the digestion of grass silage by growing cattle. *British Journal of Nutrition* 63, 489–502.

Beever, D.E., Cammell, S.B., Sutton, J.D., Rowe, N. and Perrott, G.E. (1998) Energy metabolism in high yielding dairy cows. *Proceedings of the British Society of Animal Science, 1998*, Abstract 197.

Belyea, R.L., Frost, G.R., Martz, F.A., Clark, J.L. and Forkner, L.G. (1978) Body composition of dairy cattle by potassium-40 liquid scintillation detection. *Journal of Dairy Science* 61, 206–211.

Bequette, B.J., Backwell, F.R.C., Dhanoa, M.S., Walker, A., Calder, A.G., Wray-Cahen, D., Metcalf, J.A., Sutton, J.D., Beever, D.E., Lobley, G.E. and MacRae, J.C. (1994) Kinetics of blood free and milk casein-amino acid labelling in the dairy goat at two stages of lactation. *British Journal of Nutrition* 72(8), 211–220.

Bergman, E.N. (1986) Splanchnic and peripheral uptake of amino acids in relation to the gut. *Federation Proceedings (FASEB Journal)* 45, 2277–2282.

Bickerstaffe, R., Annison, E.F. and Linzell, J.L. (1974) The metabolism of glucose, acetate, lipids and amino acids in lactating dairy cows. *Journal of Agricultural Science (Cambridge)* 82, 71–85.

Bines, J.A., Brumby, P.E., Storry, J.E., Fulford, R.J. and Braithwaite, G.D. (1978) The effect of protected lipids on nutrient intakes, blood and rumen metabolites and milk secretion in dairy cows during early lactation. *Journal of Agricultural Science (Cambridge)* 91, 135–150.

Black, J.L., Beever, D.E., Faichney, G.J., Howarth, B.R. and Graham, N.McC. (1980/81) Simulation of the effects of rumen function on the flow of nutrients from the stomach of sheep. Part 1. Description of the computer program. *Agricultural Systems* 6, 195–219.

Black, J.L., Gill, M., Beever, D.E., Thornley, J.H.M. and Oldham, J.D. (1987a) Simulation of the metabolism of absorbed energy-yielding nutrients in young sheep: efficiency of utilisation of acetate. *Journal of Nutrition* 117, 105–115.

Black, J.L., Gill, M., Thornley, J.H.M., Beever, D.E. and Oldham, J.D. (1987b) Simulation of the metabolism of absorbed energy-yielding nutrients in young sheep: efficiency of utilisation of lipid and amino acid. *Journal of Nutrition* 117, 116–128.

Blatchford, D.R. and Peaker, M. (1983) Effect of decreased feed intake on the response of milk secretion to frequent milking in goats. *Quarterly Journal of Experimental Physiology* 68, 315–318.

Blaxter, K.L. (1950) Energy feeding standards for dairy cattle. *Nutrition Abstracts and Reviews* 20, 1–21.

Blaxter, K.L. (1956) Starch equivalents, ration standards and milk production. *Proceedings of the British Society of Animal Production*, pp. 3–31.

Blaxter, K.L. (1959) Dairy cows. In: *Scientific Principles of Feeding Farm Livestock*. Proceedings of a Conference held at Brighton, November 1958. Farmer and Stockbreeder, London, pp. 21–36.

Blaxter, K.L. (1962) Progress in assessing the energy value of feeding stuffs for ruminants. *Journal of the Royal Agricultural Society* 123, 7–21.

Blaxter, K.L. (1967) The feeding of dairy cows for optimal production. George Scott Robertson Memorial Lecture. Queen's University, Belfast.

Blaxter, K.L. and Ruben, H. (1953) *First Report of the Efficiency Project (Dairy Cows)*. Report No. 239/53, Agricultural Research Council, London.

Blaxter, K.L. and Ruben, H. (1954) *Second Interim Report of the Efficiency Project (Dairy Cows)*. Report No 628/54, Agricultural Research Council, London.

Blaxter, K.L. and Wainman, F.W. (1964) The utilisation of the energy of different rations by sheep and cattle for maintenance and fattening. *Journal of Agricultural Science (Cambridge)* 63, 113–128.

Botts, R.L., Hemkin, R.W. and Bull, L.S. (1979) Protein reserves in the lactating dairy cow. *Journal of Dairy Science* 62, 433–440.

Broster, W.H. (1969) Recent research into levels of feeding for dairy cows. In: Swan, H. and Lewis, D. (eds) *Proceedings of the Third Nutrition Conference for Feed Manufacturers*. Churchill, London, pp. 53–79.

Broster, W.H. and Bines, J.A. (1974) Meeting the protein requirements of the dairy cow. *Proceedings of the British Society of Animal Production* 3, 59–67.

Broster, W.H. and Broster, V.J. (1984) Reviews of the progress of dairy science. Long-term effects of plane nutrition on the performance of the dairy cow. *Journal of Dairy Research* 51, 149–196.

Broster, W.H. and Broster, V.J. (1998) Body score of dairy cows. *Journal of Dairy Research* 65, 155–173.

Broster, W.H. and Thomas, C. (1981) The influence of level and pattern of concentrate input on milk output. In: Haresign, W. (ed.) *Recent Advances in Animal Nutrition – 1981*. Butterworths, London, pp. 49–69.

Broster, W.H., Sutton, J.D. and Bines, J.A. (1979) Concentrate: forage ratios for high yielding dairy cows. In: Haresign, W. and Lewis, D. (eds) *Recent Advances in Animal Nutrition – 1978*. Butterworths, London, pp. 99–126.

Broster, W.H., Sutton, J.D., Bines, J.A., Broster, V.J., Smith, T., Siviter, J.W., Johnson, V.W., Napper, D.J. and Schuller, E. (1985) The influence of plane of nutrition and diet composition on the performance of dairy cows. *Journal of Agricultural Science (Cambridge)* 104, 535–557.

Bruce, J.M., Broadbent, P.J. and Topps, J.H. (1984) A model of the energy system of lactating and pregnant cows. *Animal Production* 38, 351–362.

Brumby, P.E., Storry, J.E., Sutton, J.D. and Oldham, J.D. (1979) Fatty acid digestion and utilization by dairy cows. *Annales de Recherches Veterinaires* 10, 317–319.

Burt, A.W.A. (1957) The influence of level of feeding during lactation upon the yield and composition of milk. *Dairy Science Abstracts* 19, 435–454.

Butler-Hogg, B.W., Wood, J.D. and Bines, J.A. (1985) Fat partitioning in British Friesian cows: the influence of physiological state on dissected body composition. *Journal of Agricultural Science (Cambridge)* 104, 519–528.

Buttery, P.J. (1977) Aspects of the biochemistry of rumen fermentation and their implication in ruminant productivity. In: Haresign, W. and Lewis, D. (eds) *Recent Advances in Ruminant Nutrition – 1977*. Butterworths, London, pp. 8–24.

Chalmers, J.S., Veerkamp, R.F., Parkinson, H., McGinn, R., Simm, G. and Oldham, J.D. (1997) Genotype by diet and lactation interactions for yield, dry matter intake, condition score and live weight in dairy cows. *Proceedings of the British Society of Animal Science, 1997*, Abstract 39.

Chamberlain, D.G., and Thomas, P.C. (1982) Effects of intra-venous supplements of L-methionine on milk yield and composition in cows given silage-cereal diets. *Journal of Dairy Research* 49, 25–28.

Chigaru, P.R.N. and Topps, J.H. (1981) The composition of body-weight changes in underfed lactating beef cows. *Animal Production* 32, 95–103.

Childs, C.M. (1920) *Biological Bulletin of Woods Hole* 39, 147. (Cited by Hammond (1944).)

Chilliard, Y., Remond, B., Agabriel, J., Robelin, J. and Verite, R. (1987) Variations du contenu digestif et des reserves corporelles au cours du cycle gestation–lactation. *Bulletin Technique No. 70 – Decembre 1987*, 117–131.

Christie, H., Emmans, G.E., Oldham, J.D and Roberts, J.D. (1998) The effects of energy intake on response to metabolisable protein in dairy cow. *Proceedings of the British Society of Animal Science, 1997*, Abstract 20. BSAS, Penicuik, UK.

Coppock, C.E., Flatt, W.P., Moore, L.A. and Stewart, W.E. (1964) Relationship between end products of rumen fermentation and utilisation of metabolisable energy for milk production. *Journal of Dairy Science* 47 (12), 1359–1364.

Cowan, R.T., Robinson, J.J., McDonald, I. and Smart, R. (1980) Effects on body fatness at lambing and diet in lactation on body tissue loss, feed intake and milk yield of ewes in early lactation. *Journal of Agricultural Science (Cambridge)* 95, 497–514.

Crabtree, R.M. (1984) Milk compositional ranges and trends. In: Castle, M.E. and Gunn, R.G. (eds) *Milk Composition Quality and its Importance in Future Markets*. British Society of Animal Production Occasional Publication No 9, pp. 35–42.

Danfaer, A., Thysen, I. and Vagn Østergaard, (1981) Proteinnveauets indflydelse på malkekøernes produktion. I. Mœlkeydelse, tilvaekst og sundhed. *Beretning fra Statens Husdyrbrugs forsøg No 492*. Trykt Frederiksberg Bogtrykkeri, Copenhagen.

Dawson, J.M., Bruce, C.I., Buttery, P.J., Gill, M. and Beever, D.E. (1988) Protein metabolism in the rumen of silage-fed steers: effect of fishmeal supplementation. *British Journal of Nutrition* 60, 339–353.

Dewhurst, R.J. and Knight, C.H. (1993) An investigation of the changes in sites of milk storage in the bovine udder over two lactation cycles. *Animal Science* 57, 379–384.

Dijkstra, J. (1994) Simulation of the dynamics of protozoa in the rumen. *British Journal of Nutrition* 72, 679–699.

Dijkstra, J. and France, J. (1996) A comparative evaluation of models of whole rumen function. *Annales de Zootechnie* 45 (Suppl.), 175–192.

Dijkstra, J., Neal, H.D.St C, Beever, D.E. and France, J. (1992) Simulation of nutrient digestion, absorption and outflow rate in the rumen: model description. *Journal of Nutrition* 122, 2239–2256.

Dijkstra, J., Boer, H., van Bruchem, J., Bruining, M. and Tamminga S. (1993) Absorption of volatile fatty acids from the rumen of lactating dairy cows as influenced by volatile fatty acid concentration, pH and rumen liquid volume. *British Journal of Nutrition* 69, 385–396.

Dijkstra, J., France, J. and Tamminga, S. (1998) Integration of *in vivo* parameters by mechanistic modelling to predict recycling of microbial nitrogen in the rumen. In: Deaville, E.R., Owen, E., Rymer, C., Adesogan, A.T., Huntington, J.A. and Lawrence, T.L.J. (eds) *Proceedings of BSAS International Symposium, In vitro Techniques for Measuring Nutrient Supply to Ruminants, Reading*, 8–10 July. British Society of Animal Science Occasional Publication No. 22, in press.

Dils, R. (1983) Milk fat synthesis. In: Mepham, T.B. (ed.) *Biochemistry of Lactation*. Elsevier, New York, pp. 141–157.

Dodd, F.H. (1984) Herd management effects on compositional quality. In: Castle, M.E. and Gunn, R.G. (eds) *Milk Compositional Quality and its Importance in Future Markets*. British Society of Animal Production Occasional Publication No 9, pp. 77–83.

Durand, D., Bauchart, D., Lefaivre, J. and Donnat, J.P. (1988) Method for continuous measurement of blood metabolite hepatic balance in conscious pre-ruminant calves. *Journal of Dairy Science* 71, 1632–1637.

Emmans, G.C. and Fisher, C. (1986) Problems in nutritional theory. In: Fisher, C. and Boorman, K.N. (eds) *Nutrient Requirements of Poultry and Nutritional Research*. Butterworths, London, pp. 9–39.

Emmans, G.C. and Friggens, N.C. (1995) The effects of live weight, and of the quality of the feed given previously, on the *ad libitum* intakes of a poor quality hay by sheep of 5 breeds of different mature live weights. *Annales de Zootechnie* 44 (Suppl.), 269.

Evans, R.E. (1960) *Rations for Livestock*. Ministry of Agriculture Bulletin No. 48, HMSO, London.

Evans, R.A.S., Forbes, J.M. and Johnson, C.L. (1985) A commercially applicable simulation of the lactating cow. *Animal Production* 40, 560 (Abstract no. 142).

Faichney, G.J. (1975) The use of markers to partition digestion within the gastro-intestinal tract of ruminants. In: McDonald, I.W. and Warner, A.C.I. (eds) *Physiology of Digestion in the Ruminant*. University of New England, Armidale, pp. 277–291.

Fisher, C., Morris, T.R. and Jennings, R.C. (1973) A model for the description and prediction of the response of laying hens to amino acid intake. *British Poultry Science* 14, 469–484.

Fox, D.G., Sniffen, C.J. and O'Connor, J.D. (1988) Adjusting nutrient requirements of beef cattle for animal and environmental variations. *Journal of Animal Science* 66, 1475–1495.

Fox, D.G., Sniffen, C.J., O'Connor, J.D., Russell, J.B. and Van Soest, P.J. (1992) A net carbohydrate and protein system for evaluating cattle diets: III. Cattle requirements and diet adequacy. *Journal of Animal Science* 70, 3578–3596.

France, J. and Thornley, J.H.M. (1984) *Mathematical Models in Agriculture*. Butterworths, London.

France, J., Thornley, J.H.M. and Beever, D.E. (1982) A mathematical model of the rumen. *Journal of Agricultural Science (Cambridge)* 99, 343–353.

Friggens, N.C., Emmans, G.C. and Veerkamp, R.F. (1998) On the use of simple ratios between lactation curve coefficients to describe parity effects on potential milk production. *Livestock Production Science*, in press.

Fullerton, F.M., Fleet, I.R., Heap, R.B., Hart, I.C. and Mepham, T.B. (1989) Cardiovascular responses and mammary uptake in Jersey cows treated with pituitary derived growth hormone during late lactation. *Journal of Dairy Research* 56 (1), 27–35.

Garnsworthy, P.C. (1988) The effect of energy reserves at calving on performance of dairy cows. In: Garnsworthy, P.C. (ed.) *Nutrition and Lactation in the Dairy Cow*. Butterworths, London, pp. 157–170.

Garnsworthy, P.C. and Topps, J.H. (1982) The effect of body condition of dairy cows at calving on their food intake and performance when given complete diets. *Animal Production* 35, 113–120.

Gibb, M.J. and Ivings, W.E. (1993) A note on the estimation of the body fat, protein and energy content of lactating Holstein–Friesian cows by measurement of condition score and live-weight. *Animal Production* 56, 281–283.

Gibb, M.J., Ivings, W.E., Dhanoa, M.S. and Sutton, J.D. (1992) Changes in body components of autumn-calving Holstein–Friesian cows over the first 29 weeks of lactation. *Animal Production* 55, 339–360.

Gill, M., Thornley, J.H.M., Black, J.L., Oldham, J.D. and Beever, D.E. (1984) Simulation of the metabolism of absorbed energy-yielding nutrients in young sheep. *British Journal of Nutrition* 52, 621–649.

Girard, J. and Ferre, P. (1982) Metabolic and hormonal changes around birth. In: Jones, C.T. (ed.) *The Biochemical Development of the Fetus and Neonate*. Elsevier Biomedical Press, Amsterdam, pp. 517–541.

Givens, D.I. (1994) Future forage characterisation scheme. In: *Characterisation of Feeds for Farm Animals*. British Society of Animal Production Workshop Publication No. 1, BSAP, Penicuik, Midlothian, p. 41.

Goering, H.K. and Van Soest, P.J. (1970) *Forage Fiber Analysis*. Agricultural Handbook No. 379, Agricultural Research Service, USDA, Washington, DC.

Gordon, F.J. (1979) The effect of protein content of the supplement for dairy cows with access *ad libitum* to high digestibility, wilted grass silage. *Animal Production* 28, 183–189.

Graham, N.McM., Black, J.L., Faichney, G.J. and Arnold, G.W. (1976) Simulation of growth and production in sheep. Model 1. A computer programme to estimate energy and nitrogen utilization, body composition and empty live weight change day by day for sheep of any age. *Agricultural Systems* 1, 113–138.

Hammond, J. (1944) Physiological factors affecting birth weight. *Proceedings of the Nutrition Society* 2, 8–12.

Hart, I.C. (1983) Endocrine control of nutrient partition in lactating ruminants. *Proceedings of the Nutrition Society* 42, 181–194.

Hobson, P.N. (1969) Rumen bacteria. *Methods in Microbiology* 3B, 133–149.

Hogan, J.P. and Weston, R.H. (1970) Quantitative aspects of microbial protein. In: Phillipson, A.T. (ed.) *Physiology of Digestion and Metabolism in the Ruminant*. Oriel Press, Newcastle upon Tyne, pp. 474–485.

Hovell, F.D.DeB., MacPherson, R.M., Crofts, R.M.J. and Pennie, K. (1977) The effect of energy intake and mating weight on growth, carcass yield and litter size of female pigs. *Animal Production* 25, 233–245.

Hulme, D.J., Kellaway, R.C., Booth, P.J. and Bennet, L. (1986) The CAMDAIRY model for formulating and analysing dairy cow rations. *Agricultural Systems* 22, 81–108.

Huntington, G.B., Eisemann, J.H. and Whitt, J.M. (1990) Portal blood flow in beef steers: comparison of techniques and relation to hepatic blood flow, cardiac output and oxygen uptake. *Journal of Animal Science* 68, 1666–1673.

INRA (1978) *Alimentation des Ruminants.* INRA, Versailles.

INRA (1988) *Alimentation des Bovins, Ovins, et Caprins* (Jarrige, R., ed.). INRA, Paris.

Ivings, W.E., Gibb, M.J., Dhanoa, M.S. and Fisher, A.V. (1993) Relationships between velocity of ultrasound in live lactating dairy cows and some post-slaughter measurements of body composition. *Animal Production* 56, 9–16.

Jensen, E., Klein, J.W., Rauchenstein, E., Woodward, T.E. and Smith, R.H. (1942) *Input–output Relationships in Milk Production.* Technical Bulletin No. 815, United States Department of Agriculture, Washington, DC.

Johnson, C.L. (1973) Some aspects of changing body composition of mice during successive pregnancies and lactations. *Journal of Endocrinology* 56, 37–46.

Johnson, C.L. (1983a) Influence of feeding pattern on the biological efficiency of high-yielding dairy cows. *Journal of Agricultural Science (Cambridge)* 100, 191–199.

Johnson, C.L. (1983b) The effect of level and pattern of feeding on the yield of milk constituents from cows of different yield potential. *Journal of Agricultural Science (Cambridge)* 101, 717–726.

Johnsson, I.D., Hart, I.C. and Turvey, A. (1986) Pre-pubertal mammogenesis in the sheep. III. The effects of restricted feeding or daily administration of bovine growth hormone and bromocriptine on mammary growth and morphology. *Animal Production* 42, 53–63.

Jones, B.B., Kellaway, R.C. and Lean, I.J. (1996) *Protein Requirements of Dairy Cows.* Dairy Research and Development Corporation, Glen Iris, Australia.

Kellner, O. (1905) *Die Ernahrung der Landwirtschaftlichen Nutziere Verlays-buchhandlung.* Paul Parey, Berlin.

Komarek, R.J. (1981) Intestinal cannulation of cattle and sheep with a T-shaped cannula designed for total digesta collection without externalising digesta flow. *Journal of Animal Science* 53, 796–802.

Konig, B.A., Parker, D.S. and Oldham, J.D. (1979) Acetate and palmitate kinetics in lactating dairy cows. *Annales des Recherche Veterinaire* 10, 368–370.

Land, C. and Leaver, J.D. (1980) The effect of body condition at calving on the milk production and feeding intake of dairy cows. *Animal Production* 30, 449 (Abstract no. 4).

Lees, J.A., Oldham, J.D., Haresign, W. and Garnsworthy, P.C. (1990) The effect of patterns of rumen fermentation on the response by dairy cows to dietary protein concentration. *British Journal of Nutrition,* 63 177–186.

Linzell, J.L. (1974) Mammary blood flow and methods of identifying and measuring precursors in milk. In: Larson, B.L. and Smith, V.R. (eds) *Lactation,* Vol. 1. Academic Press, New York and London, pp. 143–225.

Maas, J.A., Donovan, K.C. and Baldwin, R.L. (1997) A user-friendly interface for dynamic computer models of dairy cow metabolism for teaching and extension. *Journal of Dairy Science* 80 (Suppl.1), P238.

McAllan, A.B., Knight, R. and Sutton, J.D. (1983) The effect of free and protected oils on the digestion of dietary carbohydrates between the mouth and duodenum of sheep. *British Journal of Nutrition* 49, 433–440.

McAllan, A.B., Siddons, R.C. and Beever, D.E. (1987) The efficiency of conversion of degraded nitrogen to microbial nitrogen in the rumen of sheep and cattle. In: Jarrige, R. and Alderman, G. (eds) *Feed Evaluation and Protein Requirement Systems*. CEC, Luxembourg, pp. 111–128.

MacRae, J.C. (1975) The use of re-entrant cannulae to partition digestive function within the gastro-intestinal tract of ruminants. In: McDonald, I.W. and Warner, A.C.I. (eds) *Physiology of Digestion in the Ruminant*. University of New England, Armidale, pp. 261–276.

MacRae, J.C. and Beever, D.E. (1997) Predicting amino acid supply and utilisation in the lactating ruminant. *Proceedings of the Society of Nutrition and Physiology* 6, 15–30.

MacRae, J.C., Buttery, P.J. and Beever, D.E. (1988) Nutrient interactions in the dairy cow. In: Garnsworthy, P.C. (ed.) *Nutrition and Lactation in the Dairy Cow*. Butterworths, London, pp. 55–75.

MacRae, J.C., Bruce, L.A., Hovell, F.D.deB., Hart, I.C., Inkster, J., Walker, A. and Atkinson, T. (1991) Influence of protein nutrition on the response of growing lambs to exogenous bovine growth hormone. *Journal of Endocrinology* 130, 53–61.

Madsen, J. (1985) The basis for the Nordic protein evaluation system for ruminants. The AAT/PBV system. *Acta Agriculturae Scandinavica*, Suppl. 25, 9–20.

MAFF (1921) *Rations for Livestock*, 1st Edn. Bulletin No. 48, Ministry of Agriculture and Fisheries, HMSO, London.

MAFF (1960) *Rations for Livestock*, 15th Edn. Bulletin No. 48, Ministry of Agriculture, Fisheries and Food, HMSO, London.

MAFF (1972) Report of Working Party on Energy. In: *Joint Conference on Nutrient Standards for Ruminants*. Internal Report, MAFF/ADAS, London, pp. 20–84.

MAFF (1975) *Energy Allowances and Feeding Systems for Ruminants*. Technical Bulletin 33, Ministry of Agriculture Fisheries and Food, HMSO, London.

Mepham, T.B. (1982) Amino acid utilisation by lactating mammary gland. *Journal of Dairy Science* 65, 287–298.

Mepham, T.B. (1987) Nutrient uptake in the mammary gland. In: Garnsworthy, P.C. (ed.) *Nutrition and Lactation in the Dairy Cow*. Butterworths, London, pp. 15–31.

Mertens, D.R. (1985) Effect of fiber on feed quality for dairy cows. In: *48th Minnesota Nutrition Conference*. St Paul, Minnesota, p. 209.

Metcalf, J.A., Beever, D.E., Sutton, J.D., Wray-Cahen, D., Evans, R.T., Humphries, D.J., Backwell, F.R.C., Bequette, B.J. and MacRae, J.C. (1994) The effect of supplementing protein on *in vivo* metabolism of the mammary gland in lactating dairy cows. *Journal of Dairy Science* 77, 1816–1827.

Metcalf, J.A., Wray-Cahen, D., Chettle, E.E., Sutton, J.D., Beever, D.E., Crompton, L.A., MacRae, J.C., Bequette, B.J. and Backwell, F.R.C. (1996a) The effect of increasing levels of dietary crude protein as protected soybean meal on mammary metabolism in the lactating dairy cow. *Journal of Dairy Science* 79, 603–611.

Metcalf, J.A., Crompton, L.A., Backwell, F.R.C., Bequette, B.J., Lomax, M.A., Sutton, J.D., MacRae, J.C. and Beever, D.E. (1996b) The response of dairy cows to intravascular administration of two mixtures of amino acids. *Animal Science* 62 (3), 643, Abstract 88.

Milligan, L.P. and Summers, M. (1986) The biological basis of maintenance and its relevance to assessing responses to nutrients. *Proceedings of the Nutrition Society* 45, 185–193.

Moe, P.W., Flatt, W.P. and Tyrrell, H.F. (1972) Net energy value of feeds for lactation. *Journal of Dairy Science* 55, 945–958.

Moore, J.H. and Christie, W.W. (1979) Lipid metabolism in the mammary gland of ruminants. *Progress in Lipid Research* 17, 347–395.

Murphy, M.R., Baldwin, R.L. and Koong, L.J. (1982) Estimation of stoichiometric parameters for rumen fermentation of roughage and concentrate diets. *Journal of Animal Science* 55, 411–421.

Neal, H.D.StC. and Thornley, J.H.M. (1983) The lactation curve in cattle: a mathematical model of the mammary gland. *Journal of Agricultural Science (Cambridge)* 101, 389– 400.

Nocek, J.E. and Tamminga, S. (1991) Site of digestion of starch in the gastrointestinal tract of dairy cows and its effect on milk yield and composition. *Journal of Dairy Science* 74, 3598–3629.

NRC (1985) *Ruminant Nitrogen Usage.* US National Academy of Science, Washington, DC.

NRC (1988) *Nutrient Requirements of Dairy Cattle*, 6th Edn. US National Academy of Science, Washington, DC.

O'Connor, J.D., Sniffen, C.J., Fox D.G. and Chalupa, W. (1993) A net carbohydrate and protein system for evaluating cattle diets: IV. Predicting amino acid adequacy. *Journal of Animal Science* 71, 1298–1311.

Oldham, J.D. (1984) Protein–energy interrelationships in dairy cows. *Journal of Dairy Science* 67, 1090–1114.

Oldham, J.D. (1987) Efficiencies of amino acid utilisation. In: Jarrige, R. and Alderman, G. (eds) *Feed Evaluation and Protein Requirement Systems for Ruminants.* CEC, Luxembourg, pp. 171–186.

Oldham, J.D. (1996) Will responses to protein become more predictable? Proceedings of Society of Chemical Industry, Agriculture and Environment Group Symposium: the Scope for the Future Development of the UK Metabolisable Protein System. *Journal of Science of Food and Agriculture*, SCI Lecture Paper Series, Paper No. 0012.

Oldham, J.D. and Emmans, G.C. (1988) Prediction of responses to protein and energy yielding nutrients. In: Garnsworthy, P.C. (ed.) *Nutrition and Lactation in the Dairy Cow.* Butterworths, London, pp. 76–96.

Oldham, J.D. and Emmans, G.C. (1989a) Animal performance as the criterion for feed evaluation. In: Cole, D.J.A. and Wiseman, J.S. (eds) *Feedstuff Evaluation.* Butterworths, Guildford, pp. 73–90.

Oldham, J.D. and Emmans, G.C. (1989b) Prediction of responses to required nutrients in dairy cows. *Journal of Dairy Science* 72, 3212–3229.

Oldham, J.D., Phipps, R.J., Fulford, R.J., Napper, D.J., Thomas, J. and Weller, R.F. (1985) Responses of dairy cows to rations varying in fishmeal or soyabean meal content in early lactation. *Animal Production* 40, 519 (Abstract no. 3)

Ørskov, E.R. and McDonald, I. (1979) The estimation of protein degradability in the rumen from incubation measurements weighted according to rate of passage. *Journal of Agricultural Science (Cambridge)* 92, 499–503.

Ørskov, E.R., Flatt, W.P., Moe, P.W., Munson, A.W., Hemken, R.W. and Katz, I. (1969) The influence of ruminal infusion of volatile fatty acids on milk yield and composition and on energy utilization by lactating cows. *British Journal of Nutrition* 23, 443–453.

Ørskov, E.R., Grubb, D.A. and Kay, R.N.B. (1977) Effect of post-ruminal glucose or protein supplementation on milk yield and composition in Friesian cows in early lactation and negative energy balance. *British Journal of Nutrition* 38, 397–405.

Pine, A.P., Jessop, N.S. and Oldham, J.D. (1994) Maternal protein reserves and their influence on lactational performance in rats. *British Journal of Nutrition* 71, 13–27.

Pirt, S.J. (1965) The maintenance energy of bacteria in growing cultures. *Proceedings of the Royal Society of London, Series B. Biological Science* 163, 224–231.

Pitt, R.E., Van Kessel, J.S., Fox, D.G., Pell, A.N., Barry, M.C. and Van Soest, P.J. (1996) Prediction of ruminal volatile fatty acids and pH within the net carbohydrate and protein system. *Journal of Animal Science* 74, 226–244.

Prewitt, L.R., Jacobson, D.R., Hemken, R.W. and Hatton, R.H. (1975) Portal blood flow as measured by the Doppler shift technique in sheep fed chopped forages. *Journal of Animal Science* 41, 596–600.

Reeds, P.J., Fuller, M.F., Cadenhead, A. and Hay, A.M. (1987) Urea synthesis and leucine turnover in growing pigs: changes during 2d following the addition of carbohydrate or fat to the diet. *British Journal of Nutrition* 58, 301–311.

Reeds, P.J. and Fiorotto, M.L. (1990) Growth in perspective. *Proceedings of the Nutrition Society* 49, 411–420.

Reid, J.T. and Robb, J. (1971) Relationships of body composition to energy intake and energetic efficiency. *Journal of Dairy Science* 54, 553–564.

Reid, I.M., Roberts, C.J., Treacher, R.J. and Williams, L.A. (1986) Effect of body condition at calving on tissue mobilization, development of fatty liver and blood chemistry of dairy cows. *Animal Production* 43, 7–15.

Reynolds, C.K. (1995) Quantitative aspects of liver metabolism in ruminants. In: Engelhardt, W.V., Leonhard-Marek, S., Breves, G. and Gisecke, D. (eds) *Ruminant Physiology: Digestion, Metabolism, Growth and Reproduction, Proceedings 8th International Symposium on Ruminant Physiology*. Delmar Publishers, Albany, Germany, pp. 351–371.

Reynolds, C.K., Harmon, D.L. and Cecavu, M.J. (1994) Absorption and delivery of nutrients for milk protein synthesis by portal-drained viscera. *Journal of Dairy Science* 77, 2787–2808.

Robinson, J.J. (1986) Changes in body composition during pregnancy and lactation. *Proceedings of the Nutrition Society* 45, 71–80.

Roe, W.E., Bergman, E.N. and Kon, K. (1966) Absorption of ketone bodies and other metabolites via the portal blood of sheep. *American Journal of Veterinary Research* 27, 729–736.

Rook, A.J.F. (1961) Variations in the chemical composition of the milk of the cow. *Dairy Science Abstracts* 23, 251–258, 303–308.

Rook, A.J.F. and Thomas, P.C. (1983) Milk secretion and its nutritional regulation.

In: Rook, A.J.F. and Thomas, P.C. (eds) *Nutritional Physiology of Farm Animals*. Longman, London, pp. 314–368.

Rulquin, H., Pisulewski, P.M., Verité, R. and Guinard, J. (1993) Milk production and composition as a function of post-ruminal lysine and methionine supply: a nutrient-response based approach. *Livestock Production Science* 37, 69–90.

Russell, J.B., O'Connor, J.D., Fox, D.G., Van Soest, P.J. and Sniffen, C.J. (1992) A net carbohydrate and protein system for evaluating cattle diets: I. Ruminal fermentation. *Journal of Animal Science* 70, 3551–3561.

Sano, H., Hayakawa, S., Takahashi, H. and Terashima, Y. (1995) Plasma insulin and glucagon responses to propionate infusion into femoral and mesenteric veins in sheep. *Journal of Animal Science* 73, 191–197.

SCA (1990) *Feeding Standards for Australian Livestock*. Standing Committee of Agriculture, Ruminants Sub-Committee, CSIRO Publications, Melbourne.

Seal, C.J. and Reynolds, C.K. (1993) Nutritional implications of gastrointestinal and liver metabolism in ruminants. *Nutrition Research Reviews* 6, 185–208.

Serjsen, K., Faldager, J., Sorensen, M.T. and Bauman, D.E. (1983) Regulation of mammary development in dairy heifers. In: *Proceedings of the 34th Annual Meeting of the European Association of Animal Production, Madrid*. Paper No. C5a.2, Ministerio de Agricultura, Pesca y Alimentation, Secretaria General Technica, Monpre, S.L., Spain.

Smith, N.E. and Baldwin, R.L. (1974) Effects of breed, pregnancy and lactation on weight of organs and tissues in dairy cattle. *Journal of Dairy Science* 57, 1055–1060.

Smith, G.H., Crabtree, B. and Smith, R.A. (1983) Energy metabolism in the mammary gland. In: Mepham, T.B. (ed.) *Biochemistry of Lactation*. Elsevier, New York, pp. 121–140.

Smith, G.H. and Taylor, D.S. (1977) Mammary energy metabolism. In: Peaker, M. (ed.) *Comparative Aspects of Lactation*. Academic Press, London, pp. 95–109.

Sniffen, C.J., O'Connor, J.D., Van Soest, P.J., Fox, D.G. and Russell, J.B. (1992) A net carbohydrate and protein system for evaluating cattle diets: II. Carbohydrate and protein availability. *Journal of Animal Science* 70, 3562–3577.

Storry, J.E., Hall, A.J., Tuckley, B. and Millard, D. (1969) The effects of intravenous infusions of cod-liver and soya-bean oils on the secretion of milk fat in the cow. *British Journal of Nutrition* 23, 173–180.

Strickland, M.J. and Poole, D.A. (1985) An evaluation of protein levels and sources in concentrates for dairy cows. *Animal Production* 40, 519 (Abstract no. 1).

Subnel, A.P.J., Meijer, R.G.M., van Straalen, W.M. and Tamminga, S. (1994) Efficiency of milk production in the DVE protein evaluation system. *Livestock Production Science* 40, 215–224.

Sutton, J.D. (1985) Digestion and absorption of energy substrates in the lactating cow. *Journal of Dairy Science* 68, 3376–3393.

Sutton, J.D. (1986) Milk composition. In: Broster, W.H., Phipps, R.H. and Johnson, C.L. (eds) *Principles and Practice of Feeding Dairy Cows*. Technical Bulletin No. 8, National Institute for Research in Dairying, Reading, pp. 203–218.

Sutton, J.D., Oldham, J.D. and Hart, I.C. (1980) Products of digestion, hormones and energy utilization in milking cows given concentrates containing varying proportions of barley or maize. In: Mount, L.E. (ed.) *Energy Metabolism*. Butterworths, London, pp. 303–306.

Sutton, J.D., Knight, R., McAllan, A.B. and Smith, R.H. (1983) Digestion and synthesis in the rumen of sheep given diets supplemented with free and protected oils. *British Journal of Nutrition* 49, 419–432.

Sutton, J.D., Bines, J.A., Morant, S.V., Napper, D.J. and Givens, D.I. (1987) A comparison of starchy and fibrous concentrates for milk production, energy utilization and hay intake by Friesian cows. *Journal of Agricultural Science (Cambridge)* 109, 375–386.

Sykes, A.R. and Field, A.C. (1974) Seasonal changes in plasma concentrations of proteins, urea, glucose, calcium and phosphorus in sheep grazing a hill pasture and their relationship to changes in body composition. *Journal of Agricultural Science (Cambridge)* 83, 161–169.

Tamminga, S., Van Straalen, W.M., Subnel, A.P.J., Meijer, R.G.M., Steg, A., Wever, C.J.G. and Blok, M.C. (1994) The Dutch protein evaluation system: the DVE/OEB-system. *Livestock Production Science* 40, 139–155.

Taylor, St C.S. (1973) Genetic differences in milk production in relation to mature body weight. *Proceedings of the British Society of Animal Production* 2, 15–26.

Taylor, St C.S. (1985) The use of genetic size-scaling in the evaluation of animal growth. *Journal of Animal Science* 61 (Suppl. 2), 118–143.

Theodorou, M.K., Williams, B.A., Dhanoa, M.S., McAllan, A.B and France, J. (1994) A simple gas production method using a pressure transducer to determine the fermentation kinetics of ruminant feeds. *Animal Feed Science and Technology* 48, 185–197.

Thomas, C. (1985) Milk compositional quality and the role of forages. In: Castle, M.E. and Gunn, R.G. (eds) *Milk Compositional Quality and its Importance in Future Markets*. British Society of Animal Production Occasional Publication No. 9, pp. 69–76.

Thomas, C., Aston, K., Daley, S.R. and Beever, D.E. (1985) The effect of level and pattern of protein supplementation on milk output. *Animal Production* 40, 519 (Abstract no. 2).

Thomas, P.C. and Chamberlain, D.G. (1984) Manipulation of milk composition to meet market needs. In: Haresign, W. and Cole, D.J.A. (eds) *Recent Advances in Animal Nutrition – 1984*. Butterworths, London, pp. 219–243.

Trayhurn, P. (1996) New insights into the development of obesity: obese genes and the leptin system. *Proceedings of the Nutrition Society* 55, 783–791.

Tyrrell, H. (1980) Limits to milk production efficiency. *Journal of Animal Science* 51 (6), 1441–1447.

Ulyatt, M.J. and Egan, A.R. (1979) Quantitative digestion of fresh herbage by sheep. V. The digestion of four herbages and prediction of sites of digestion. *Journal of Agricultural Science (Cambridge)* 92, 605–616.

Van der Honing, Y. (1975) *Intake and Utilisation of Energy with Rations with Pelleted Forages by Dairy Cows*. Wageningen Agricultural Research Reports 836, Centre for Agricultural Publishing and Documentation, Wageningen.

Van Es, A.J.H. (1978) Feed evaluation for ruminants. I. The system in use from May, 1977 onwards in The Netherlands. *Livestock Production Science* 5, 331–345.

Van Es, A.J.H., Nijkamp, H.J. and Vogt, J.E. (1970) Feed evaluation in dairy cows. In: Schurch, A. and Wenk, C. (eds) *Proceedings of the 5th Energy Metabolism Symposium*. EAAP Publication No.13, Juris Druck Verlag, Zurich, pp. 61–64.

Van Straalen, W.M. and Tamminga, S. (1990) Protein degradation in ruminant diets.

In: Wiseman, J. and Cole, D.J.A. (eds) *Feedstuff Evaluation*. Butterworths, Sevenoaks, pp. 55–72.

Veerkamp, R.F., Simm, G. and Oldham, J.D. (1994) Effects of interaction between genotype and feeding system on milk production, feed intake, efficiency and body tissue mobilisation in dairy cows. *Livestock Production Science* 39, 229–241.

Vernon, R.G. (1981) Lipid metabolism in the adipose tissue of ruminant animals. In: Christie, W.W. (ed.) *Lipid Metabolism in Ruminant Animals*. Pergamon Press, Oxford, pp. 279–362.

Waite, R., White, J.C.D. and Robertson, A. (1956) Variations in the chemical composition of milk with particular reference to the solids-not-fat. I. The effect of stage of lactation, season of year and age of cow. *Journal of Dairy Research* 23, 65–81.

Walsh, A., Sutton, J.D. and Beever, D.E. (1992) Body composition and performance of autumn-calving Holstein–Friesian dairy cows during lactation: metabolic activity *in vitro* of subcutaneous adipose tissue. *Animal Production* 54 (3), 475, Abstract 104.

Wangsness, P.J. and McGilliard, A.D. (1972) Measurement of portal blood flow in calves by dye-dilution. *Journal of Dairy Science* 55, 1439–1446.

Webster, A.J.F. (1978) Measurement and prediction of methane production, fermentation heat and metabolism in the tissues of the ruminant gut. In: Osbourn, D.F., Beever, D.E. and Thomson, D.J. (eds) *Ruminant Digestion and Feed Evaluation*. ARC, London, pp. 8.1–8.10.

Webster, A.J.F. (1992) The metabolizable protein system for ruminants. In: Garnsworthy, P., Haresign, W. and Cole, D.J.A. (eds) *Recent Advances in Animal Nutrition – 1992*. Butterworth Heinemann, Oxford, pp. 93–110.

Webster, A.J.F., Kitcherside, M.A., Kierby, J. and Hall, P.A. (1984) Evaluation of protein feeds for dairy cows. *Animal Production* 38, 548 (Abstract no. 165).

Westgarth, D.R., Phipps, R.H., Weller, R.F. and Thomas, J. (1981) Influence of feed allocation during lactation on milk yield and quality, liveweight change and profit margin. In: *Annual Report of the National Institute for Research in Dairying, 1981*. National Institute for Research in Dairying, Shinfield, pp. 30–31.

White, S.W., Chalmers, J.P., Hilder, R. and Korner, P.I. (1967) Local thermo-dilution method for measuring blood flow in the portal and renal veins of the un-anaesthetised rabbit. *Australian Journal of Experimental Biology and Medical Science* 45, 453–468.

Whitelaw, F.G., Milne, J.S., Ørskov, E.R. and Smith, J.S. (1986) The nitrogen and energy metabolism of lactating cows given abomasal infusions of casein. *British Journal of Nutrition* 55, 537–556.

Wilson, P.N. and Strachan, P.J. (1981) The contribution of undegraded protein to the protein requirements of dairy cows. In: Haresign, W. (ed.) *Recent Advances in Animal Nutrition – 1980*. Butterworths, London, pp. 99–118.

Wood, P.D.P. (1976) Algebraic models of the lactation curves for milk, fat and protein production, with estimates of seasonal variation. *Animal Production* 22, 35–40.

Wood, P.D.P. (1979) A simple model of lactation curves for milk yield, food requirement and body weight. *Animal Production* 28, 55–63.

Wright, I. and Russell, A.J.F. (1984) The composition and energy content of empty body weight change in mature cattle. *Animal Production* 39, 365–369.

Yates, F., Boyd, D.A. and Pettit, G.H.N. (1942) Influence of changes in level of feeding on milk production. *Journal of Agricultural Science (Cambridge)* 32, 428–456.

Zhang, Y., Proenca, R., Maffei, M., Barone, M., Leopold, L. and Friedman, J.M. (1994) Positional cloning of the mouse obese gene and its human homologue. *Nature (London)* 372, 425–432.

Zinn, R.A., Bull, L.S., Hemken, R.W., Button, F.S., Enlow, C. and Tucker, R.W. (1980) Apparatus for measuring and sub-sampling digesta in ruminants equipped with re-entrant intestinal cannulas. *Journal of Animal Science* 51, 193–201.

Appendix: Modelling nutrient supply and metabolism in the dairy cow

A.1   Background

Within this section a brief outline of a mathematical model of nutrient metabolism in the dairy cow is provided, the purpose being to indicate one possible approach to the development of a mechanistic model which is capable of predicting milk constituent output. For a more detailed understanding of the subject see France and Thornley (1984).

Once the objective of the model has been reconciled, which in this case is to predict milk constituent output, the level of representation as cited by France and Thornley (1984) must be reconciled. Any mechanistic model, by definition, aims to understand behaviour occurring at one level (i) by describing the major processes or attributes of the system occurring at the next lowest level (i–1). Within the dairy cow, it is possible to identify at least four major areas which may influence the ultimate synthesis of milk constituents, these being the regulation of feed intake (intake), the digestion of ingested nutrients and subsequent absorption of the end-products of digestion (digestion), the utilization of absorbed nutrients, including regulation of partition within the body of the animal (body metabolism), and the synthesis of milk constituents within the mammary gland (mammary metabolism). Within the context of this report, intake will not be considered further, but schematic flow diagrams for digestion, body metabolism and mammary metabolism were presented in Figs 10.2 and 10.3 (in Chapter 10).

In the following sections, these schematic representations will be described and justified, followed by a brief resumé of model construction, to include examples of nutrient degradation and assimilation and the control of nutrient interactions within both the digestion and metabolism elements of the model. No attempt to parameterize the model will be presented, but through a systematic approach to the problem it is hoped that the reader will recognize that mechanistic models can be formulated with relative ease.

The principal components required for a mathematical representation of the dairy cow are:

1. Ruminal digestion
- total starch pool;
- degradable fibre pool;
- degraded hexose pool;
- AA pool;
- ammonia pool;
- VFA pool;
- rumen microbial biomass pool.

2. Intestinal absorption

3. Body metabolism
- acetate pool;
- glucose pool;
- AA pool;
- LCFA pool.

4. Mammary gland metabolism
- acetate pool;
- glucose pool;
- LCFA pool.

A.2   Ruminal digestion

Figure 10.2 in Chapter 10 is a schematic representation of ruminant digestion, principally energy and protein metabolism within the rumen. Dietary inputs are broadly defined as eight categories, four representing the carbohydrate fraction, three representing dietary crude protein and a further component representing dietary lipid. Within the carbohydrate component, water-soluble carbohydrate is considered to include free sugars, short-chain polymers and pectin. All are considered to be ultimately degraded within the rumen and consequently contribute, according to the amount contained in the diet to the degraded hexose (DC) pool which is a state variable of the model. Starch is represented as one input, entering the degradable starch pool where, depending upon its concentration and composition, as well as microbial activity in the rumen, it may be degraded to hexose and enter the DC pool or flow out from the rumen as undegraded starch and contribute to starch flow to the small intestine. This partition between degradation to hexose or outflow will be determined by the nature of the starch, in particular its potential degradability and the actual rate of digestion.

Whilst all fibre is initially considered as one fraction, the partition between degradable fibre and undegradable fibre allows completely unavailable fibre to be treated separately, although representation as a state variable will be important in any consideration of the effect of rumen fill on voluntary feed intake. Fibre which enters the degradable fibre pool can, like starch, according to its degradation characteristics be either degraded to constitutive sugars and enter the degraded carbohydrate (DC) pool, or contribute to the undigested fibre pool and subsequently flow out of the rumen, along with any fibre that was considered to be completely unavailable.

The crude protein fraction of the diet is represented as 'true' protein and NPN, the latter accounting for all dietary nitrogen other than free AAs, peptides and protein. All NPN is assumed to contribute in stoichiometric amounts to the rumen ammonia pool whilst the protein fraction is treated in the same way as fibre, recognizing that different proteins have different susceptibilities to degradation within the rumen. Thus protein which enters the degradable protein pool can leave either by digestion to AAs (pool AA) or enter the undegraded protein pool where it ultimately passes, along with any undegradable dietary protein into the small intestine.

Whilst AAs will be used principally in the synthesis of microbial protein, part will be degraded by microbial activity to ammonia with an associated production of a carbon skeleton which is assumed to enter the DC pool as hexose. Equally ammonia will be used to support microbial protein synthesis, but in addition to ATP, this process will also require provision of a carbon skeleton from the DC pool. At the same time excess ammonia will be absorbed from the rumen, whilst a small quantity (not represented in Fig. 10.2) is known to leave the rumen via the omasum, according to the fractional outflow rate of rumen water. To furnish the microbial requirement for energy, considerable quantities of degraded carbohydrate (DC) will be fermented, yielding ATP, with an associated production of VFA and methane. The individual VFA will be produced in relation to the species diversity of the microbial population present in the rumen, which will be influenced by the nature of the diet consumed. The VFA will be absorbed, methane will be removed primarily by eructation and ATP will contribute either to the maintenance of the microbial biomass, or the major synthetic processes occurring within the microbial biomass, with the requirements being in accord with established stoichiometric relationships. In situations where DC supply is excessive, a significant portion may be incorporated directly into microbial polysaccharides, whilst any remainder is likely to be fermented with an associated futile production of ATP (heat).

In many respects, the dietary lipid component is considered as inert to rumen fermentation. The major transformations will include hydrolysis and partial hydrogenation and thus the need to include a state variable for lipid is only necessary if hydrogen transactions are to be accounted for, or if microbial lysis, with the release of microbial lipid is represented. Clearly, inclusion of this latter aspect would necessitate a more detailed representation to include the disposal of microbial polysaccharides, protein and nucleic acids.

In summary, the nutrients which flow into the small intestine will comprise undigested fibre, starch and protein, ruminally modified dietary lipid and the microbial component. Thereafter digestion within the small intestine and caecum and colon is empirically represented according to previously derived estimates of the apparent digestibilities of such components within the post-duodenal section of the alimentary tract.

A.2.1   Model construction

The majority of nutrient pools in the rumen are described by assembling defined parameters for mathematical modules, most of which have the following components:

- Nutrient input calculated from dry matter intake and nutrient content of the feed.
- Nutrient concentration in the rumen, which requires information on rumen volume (l), and rumen solid and liquid turnover rates.
- A Michaelis–Menten type of equation predicting the rate of utilization of each nutrient. This equation has the form:

$$\text{flux of nutrient } A = kA/(K + A)$$

where k is the affinity constant for nutrient A, K is the maximum reaction velocity constant (V_{max}), and A is the concentration of nutrient A.

This equation requires the definition of affinity constants for the reaction in question, inhibition constants to express the effects of substrate accumulation, and maximum velocities (V_{max}) for each reaction. These components are assembled into differential equations with respect to time, which are integrated when the model is run.
- Rate constants describing the passage (flux) of individual undegraded nutrients out of the rumen into the intestines.

The characteristics of each nutrient pool will now be briefly considered:

A.2.2   Starch digestion pool and digestible fibre pool

Both the starch and digestible fibre pools follow the principles outlined above when there is only one nutrient input.

A.2.3   Digestible carbohydrate pool

The DC pool is more complex, because there are four specified inputs:

- from soluble carbohydrates (sugars) in the feed;
- from starch digestion;
- fibre digestion;
- AA deamination (described below).

These inputs are summed to give the mass of DC in the rumen and hence concentrations in rumen fluid.

There are three outputs from the DC pool:

- fermentation of DC to yield ATP and VFA;
- utilization of DC for microbial protein and nucleic acid synthesis from ammonia;
- utilization of DC for microbial polysaccharide and lipid synthesis.

Each of these reactions also follow the same basic model, requiring V_{max}, affinity and inhibition constants, plus a quantification of the amounts of DC required for each particular synthesis.

A.2.4 Amino acid pool

The amino acid pool has one input in most situations, i.e. that from the degradation of dietary protein estimated in the protein pool, dependent upon the quantity of protein consumed, the susceptibility of the ingested protein to degradation in the rumen and rumen fractional outflow rate. There are two outputs:

- Microbial protein synthesis. This requires a V_{max} and two affinity constants, one for ATP availability for biomass synthesis and one for the AA to microbial crude protein (MCP) reaction.
- Deamination of AAs.

As well as a V_{max} and an affinity constant for the reaction, an inhibition constant is also needed for ammonia concentration in the rumen liquor, which in this case operates in the opposite sense to that for microbial synthesis, i.e. high ammonia levels may inhibit deamination.

A.2.5 Ammonia pool

There are three inputs to the ammonia pool:

- dietary NPN calculated from DMI and diet composition;
- salivary NPN input requiring an assessment of salivation rate, estimated from time spent eating and ruminating as functions of intake and amount of fibre in the diet, and the NPN content of saliva;
- from AA deamination described above.

Two outputs from the ammonia pool can be defined:

- Microbial uptake of ammonia. This requires three affinity constants for ammonia to MCP, ATP requirement for biomass synthesis and DC availability.
- Ammonia absorption. This is represented by a rate constant which is a function of rumen pH and rumen ammonia concentration.

A.2.6   *Volatile fatty acid production*

VFA will be produced from the fermentation of DC described above, with the extent of fermentation being regulated in the first instance by the demands for ATP for microbial maintenance and growth. This pool is represented in the model as a zero pool (i.e. ATP can not be stored/accumulated) with one input and three outputs.

The stoichiometric yields of the principal VFA will depend upon the relative contributions of the four major substrates to the DC pool, with each primary nutrient (e.g. fibre) known to generate different VFA amounts and proportions. Clearly as the relative contributions of the primary nutrients to the DC change, so microbial interactions with respect to ATP and VFA yield will occur. At this stage, representation of this feature would be difficult, and thus the relative contributions of the primary nutrients alone will be used to determine individual VFA yields, with VFA absorption assumed to follow individual production rates. The occurrence of VFA inter-conversions within the rumen is assumed to be of little biological significance in relation to the overall objective of the model and is therefore ignored.

A.2.7   *Rumen microbial pool*

Inputs to the microbial pool will comprise microbial protein and polysaccharide (as represented in Fig. 10.2 and Sections A.2.3–A.2.5) and microbial lipid which is not included explicitly in Fig. 10.2. The composition of microbial material flowing out of the rumen will directly reflect the composition of the microbial pool appertaining at each integration step, whilst the rate of passage will be regulated by the rate of passage of water and undigested particles, thus attempting to recognize the spatial distribution of microbial biomass.

A.3   Intestinal absorption

This is not presented in a diagrammatic form and is presumed to be describable in simple terms, although it is recognized that the processes of absorption, intestinal mucosa metabolism and endogenous secretions into the alimentary tract can have a significant effect on the net gain of nutrients by the animal. Application of previously derived values for the apparent digestibility of individual nutrients within the small intestine will take account of endogenous secretions, but it may be necessary to recognize intestinal mucosa metabolism, in light of current research on portal-drained viscera metabolism which is indicating substantial utilization of glucose and AAs. Furthermore, as the overall objective at this stage is to quantify nutrient supply to peripheral tissues, some representation of metabolism within the liver may be justifiable.

In most dietary situations, the role of the caecum and colon is minimal, often failing to account for 10% of the total energy which is apparently absorbed from the whole tract. The major contributor to this is likely to be the fermentation of undigested structural carbohydrates with the principal end-products of digestion being VFA and ammonia. In the absence of more relevant data, application of the stoichiometric principles of rumen VFA production should be appropriate.

A.4   Body metabolism

The principal reactions involved in the metabolism of absorbed nutrients, after due consideration of metabolism in the portal-drained viscera and liver (see Section A.3), are represented in Fig. 10.3a in Chapter 10.

The major considerations with respect to Fig. 10.3a are the supply of nutrients and representation of nutrient partition between the body and the mammary gland. All absorbed butyrate is stoichiometrically converted to acetate which is considered as the major substrate for ATP production by oxidation and fat synthesis. Propionate, being the major precursor of glucose is explicitly represented, with provision for propionate supplied in excess of requirements for glucose synthesis to be oxidized to ATP (as represented – via acetate). Glucose will be supplied either directly from the small intestine, or via gluconeogenesis from propionate or AAs. Principal routes of glucose utilization include maintenance, lactose synthesis, provision of glycerol and reduced nicotinamide adenine dinucleotide (NADH) for fat synthesis, and oxidation with the associated production of ATP. Lipids (LCFA) are considered as major substrates for the synthesis of body and milk fat, with the intestinal supply of lipids being augmented at times of body fat mobilization. Fatty acid oxidation is explicitly represented as the production of acetate which may then be further oxidized or resynthesized into fatty acids. AAs are considered as the sole substrate for the synthesis of body and milk protein, but glucose production from AAs and AA oxidation is represented. Body protein mobilization is considered within Fig. 10.3a, although the quantitative significance of this in relation to whole body protein has not been established.

A.4.1   Acetate pool

This pool has six inputs, the three principal ones being:

- acetate and butyrate from feed digestion;
- acetate from body fat mobilization and LCFA degradation;
- acetate from AA catabolism.

There are three outputs from the acetate pool:

- acetate oxidation to ATP;
- acetate incorporation into body fat;
- acetate flux to the mammary gland for milk fat synthesis and ATP supply.

Clearly such reactions are influenced by the hormonal status of the animal and some representation of this will have to be included in the model.

A.4.2   Glucose pool

There are three principal inputs to the glucose pool:

- absorbed glucose, derived from starch digestion;
- glucose derived from the propionate pool, as described above;
- glucose derived by gluconeogenesis using AAs from the AA pool as above.

There are four outputs from this pool which support body and milk fat synthesis by the provision of NADH or glycerol phosphate:

- the flux of glucose to support body fat synthesis from acetate;
- the flux of glucose to support body fat synthesis from preformed fatty acids (lipids);
- the flux of glucose to the mammary gland;
- the oxidation of glucose to supply ATP.

A.4.3   Amino acid pool

There are two inputs to this pool:

- AAs absorbed from the gastrointestinal tract as in Section A.3;
- AAs derived from body protein catabolism.

Four outputs from the pool can be identified:

- AA flux to body protein synthesis;

- AA flux to mammary gland;
- AA flux to glucose;
- AA oxidation.

As well as V_{max}, affinity and inhibition constants for each reaction, the body protein synthesis function requires estimates of maximum achievable body protein mass and a rate of accretion which declines with approach to the maximum value for body protein mass.

A.5 Mammary gland metabolism

The principal reactions within the mammary gland in relation to maintenance and proliferation of secretory tissue and the synthesis of milk constituents are presented in Fig. 10.3b in Chapter 10. In several respects the reactions are similar to those representing body metabolism as in Fig. 10.3a, the major difference being the discrete synthesis of milk constituents and their sub-sequent removal by the process of milking. In addition to representing nutrient interactions whilst recognizing that the metabolic activity of secre-tory tissue will be influenced by the stage of lactation of the cow, the importance of milk accumulation and its feedback on secretion will need to be included.

The major precursors for milk fat synthesis will be acetate and LCFA (as lipoprotein) derived from mammary blood supply. Acetate will be the major substrate for *de novo* synthesis of fatty acids up to carbon lengths of 16, whilst a substantial amount of acetate will be oxidized to yield ATP within the mammary gland. Most of the LCFA will be incorporated directly into milk fat, whilst provision for oxidation of some LCFA is included. Requirements for glucose and ATP by both reactions are also represented.

Glucose is the sole precursor for lactose synthesis, with an accompanying ATP requirement, whilst any glucose in excess of that required to support fat synthesis will be oxidized to produce ATP. AAs will support both milk protein and tissue protein synthesis, the regulation of which will be under strict endocrinological control. Clearly both reactions require energy in the form of ATP, although the stoichiometric requirement of such may be reduced if a significant proportion of protein synthesis arises from peptide uptake. Some contribution to the AA pool will arise from mammary tissue turnover, part of this being obligatory and part related to reduction in the mass of synthetic tissue as lactation progresses.

Within the context of this whole section it is not proposed to describe the reactions associated with milk constituent synthesis in detail, largely because they possess close similarity with some of the details presented in Section A.4, and inclusion here would be largely repetitious.

A.6   Model parameterization

Once the whole model has been described in line with the principles established in Sections A.2–A.5 and illustrated in Fig. 10.3, the next step is to establish parameter values for the reaction velocities, affinity constants, inhibition constants, tissue sizes, metabolite concentrations, and stoichiometric yields and requirements. In any model building exercise this is an arduous task, requiring extensive review of the literature and collation of appropriate data. For details on how to approach this aspect, see France and Thornley (1984). It is certain that all the required values cannot be readily identified from the literature even with cautious interpolation of *in vivo* and *in vitro* data. Extrapolation of data may be necessary, but should be done with caution. Once stability of the model has been achieved, the validity of any such judgements should be examined through the extensive use of sensitivity analysis and specific experimental tests.

Index

Note: Page numbers in *italic* refer to figures and/or tables